Walter Bradlee Snow

Steam-boiler practice

In its relation to fuels and their combustion and the economic results obtained with various

methods and devices

Walter Bradlee Snow

Steam-boiler practice
In its relation to fuels and their combustion and the economic results obtained with various methods and devices

ISBN/EAN: 9783743467071

Manufactured in Europe, USA, Canada, Australia, Japa

Cover: Foto ©berggeist007 / pixelio.de

Manufactured and distributed by brebook publishing software (www.brebook.com)

Walter Bradlee Snow

Steam-boiler practice

STEAM-BOILER PRACTICE

IN ITS RELATION TO

FUELS AND THEIR COMBUSTION

AND THE

ECONOMIC RESULTS OBTAINED WITH VARIOUS METHODS AND DEVICES.

BY

WALTER B. SNOW, S.B.,

Member American Society of Mechanical Engineers.

FIRST EDITION.

FIRST THOUSAND.

NEW YORK:
JOHN WILEY & SONS.
LONDON: CHAPMAN & HALL, LIMITED.
1899.

Copyright, 1899,
BY
WALTER B. SNOW.

TO

B. H. S.

PREFACE.

STEAM-BOILER PRACTICE as here comprehended is concerned rather with the results obtained than with a detailed description of the methods and devices employed to secure the economical combustion of fuel in connection with a steam-boiler. The design and construction of the boiler, with its furnace and accessories, form the subject-matter of other works, and, therefore, do not require treatment here. Likewise special methods of operation and devices for increasing the efficiency have elsewhere received their share of attention. There appears, however, to be a place for a work simple in its treatment and reasonably comprehensive in its character, which shall deal primarily with effects rather than with causes, which shall undertake to indicate the possible gain or loss to result from a given arrangement, and shall point the way toward securing the highest efficiency in steam-boiler practice. Such a work it has been the earnest endeavor to here present; a work neither too abstruse for the practical engineer nor too rudimentary for the technical student.

As an aid to the clearer understanding of its contents, there have been presented in the introductory chapters certain related principles of pure science, a general discussion of the phenomena of combustion and steam generation, and the properties of the substances concerned. The avoidable and unavoidable losses incident to the combustion of fuel in a boiler-furnace have been collectively treated in the chapter on Efficiency of Fuels with the purpose of so emphasizing the possibilities of increased economy and indicating the relative importance of various influences as to give pertinence to the chapters which follow. In connection with the discussion of practical results it has been possible to present only brief statements as to the conditions under which they were obtained, but references have been freely given for assistance in the further study of the subject. Special attention has been given to the matter of draft and its influence upon the economic results, measured both quantitatively and commercially. In the Appendix will be found the Rules for Conducting Boiler-trials adopted by the American Society of Mechanical Engineers.

An expression of appreciation is due for the privilege of making use in this work of certain portions of the treatise on Mechanical Draft compiled by the author for the B. F. Sturtevant Company.

WATERTOWN, MASS.,
March 4, 1899.

TABLE OF CONTENTS.

CHAPTER I.
STEAM-BOILER PRACTICE.

	PAGE
Requisites for Steam-generation	2
Influence of the Character of these Requisites	2
Ultimate Efficiency of a Steam-boiler	3
Primary Cost and Fixed Charges	4
Quantitative Efficiency	5
Operating Expenses	6
Summary	6

CHAPTER II.
WATER AND STEAM.

Composition	8
Weight and Bulk	9
Expansion by Heat	11
Absolute Zero	12
Specific Heat	14
Unit of Heat	15
Mechanical Equivalent of Heat	15
Latent and Sensible Heat	17

CHAPTER III.
COMBUSTION.

Definition	24
Carbon	24
Oxygen	24
The Atomic Theory	25

TABLE OF CONTENTS.

	PAGE
Union of Carbon and Oxygen	27
Combustion of Fuel	29
Air Required for Combustion	32
Air for Dilution	35
Analysis of Flue-gases	40
Calculation of Air-supply from Gas-analysis	42
Measurement of Air-supply by Aneometer	47
Heat of Combustion	48
Calculation of the Heat of Combustion	50
Ideal Temperature of Combustion	52

CHAPTER IV.

FUELS.

Definition	57
Natural Fuels	57
Artificial Fuels	57
Wood	58
Straw and Tan	58
Bagasse	60
Peat	64
Coal	65
Lignite	68
Bituminous Coal	69
Semi-bituminous Coal	71
Semi-anthracite Coal	71
Anthracite Coal	72
Geographical Classification	73
Petroleum	73
Natural Gas	84
Artificial Fuels	84
Charcoal	84
Coke	87
Fuel-gas	88
Patent Fuels	90

CHAPTER V.

EFFICIENCY OF FUELS.

Measure of Efficiency	92
Unit of Evaporation	94
Relative Efficiency of Various Coals	96
Influence of Ash	100

TABLE OF CONTENTS. ix

	PAGE
Influence of Size of Coal	101
Influence of the Frequency of Firing	105
Loss on Account of Moisture in Coal	108
Loss on Account of Smoke	108
Loss on Account of Carbonic Oxide	110
Admission of Air above the Fire	113
Loss on Account of Excess of Air	115
Summary of Influences Affecting the Efficiency of Fuel	124
Commercial Efficiency of Coals	127
Influence of Mechanical Draft	132
Prevention of Smoke	135

CHAPTER VI.

EFFICIENCY OF STEAM-BOILERS.

Measure of Efficiency	140
Rating of Steam-boilers	143
Radiation and Convection of Furnace Heat	149
Distribution of the Heat of Combustion	151
Disposition of Heat in Steam-boilers	154
Sources of Efficiency	155
Flue Feed-water Heaters or Economizers	162
Air-heaters or Abstractors	168
Increased Tube-heating Efficiency	172
Mechanical Stokers	176
Powdered-fuel Furnaces	179

CHAPTER VII.

RATE OF COMBUSTION.

Rate of Combustion	180
Relation of Grate-surface to Heating-surface	181
Economy of High Rates of Combustion	183
Thickness of Fire	193

CHAPTER VIII.

DRAFT.

Definition	194
Relation of Pressure and Velocity	195
Efflux of Air	198
Measurement of Draft	203
Conditions of Boiler-draft	210

	PAGE
Relation of Draft and Rate of Combustion	216
Leakage of Air	225

CHAPTER IX.

CHIMNEY-DRAFT.

Principles of Chimney-draft	228
Chimney-design	231
Efficiency of Chimneys	243

CHAPTER X.

MECHANICAL DRAFT.

Definition	247
Steam-jets	247
Fans	249
Design of Fans	251
Methods of Application	262
Closed Ash-pit System	263
Closed Fire-room System	265
Induced System	267
Advantages of Mechanical Draft	270

APPENDIX.

RULES FOR CONDUCTING BOILER-TRIALS	274

STEAM-BOILER PRACTICE.

CHAPTER I.

STEAM-BOILER PRACTICE.

THE economic progress in steam-engineering which has marked the past quarter-century has been largely the result of greater refinement in the design and construction of the engine in which the steam is utilized. Future improvement along this line is to be sought principally in the increase of the steam-pressure and the expansion-ratio. But experimental results,* which, with a quadruple-expansion engine and steam of 500 pounds pressure, have already shown a steam-consumption of 9.27 pounds per hour per indicated horse-power, and a mechanical efficiency of 86.88 per cent, are indicative of the fact that we are closely approaching the maximum practical efficiency.

Until the recent past, however, the steam generator or boiler and the manner of its operation have received far less attention than they deserve. Although under the best conditions over 80 per cent of the full calorific value of the fuel may be utilized in the production of steam, yet it is true that this high standard is seldom reached in ordinary practice.† Hoadley showed an efficiency of nearly 88 per cent in his tests

---

* Obtained with Sibley College High-pressure Quadruple-expansion Experimental Engine. Trans. Amer. Soc. Mech. Engrs., vol. XVIII.

† Warm-blast Steam-boiler Furnace. J. C. Hoadley.

of a warm-blast steam-boiler furnace equipped with air-heaters and mechanical draft, while Bryan [*] has reported eighty-six tests conducted under common conditions with ordinary fuel, upon boilers of various types, which indicate an average efficiency of only 58 per cent, and have a range between a minimum of 34.6 per cent, obtained with a small vertical boiler, and a maximum of 81.32 per cent, resulting from the test of a water-tube boiler with improved setting. The possibilities of increased economy in ordinary boiler-practice are thus rendered clearly evident.

Requisites for Steam-generation.—As ordinarily considered, the primary requisites for the generation of steam are, the water from which it is transformed, the fuel by whose combustion the heat is produced, and the boiler in which the water is contained and the steam generated. To these must properly be added the air necessary for supporting combustion, while for economical generation there should also be included the means of supplying the air to the fuel, as well as the furnace in which it is burned.

Influence of the Character of these Requisites.—Nothing can be more obvious than that the character of these requisites must have a marked influence upon the efficiency with which steam is generated.

With a good quality of water and ordinary care, its character has but little effect upon the economic operation of the boiler. Such influence as it may exert when of poor quality results principally from the deposition of solid particles or the formation of scale. The limitation of this work to a discussion of boiler-practice in its relation to fuels and their combustion obviously excludes a consideration of the influence of the character of water.

Of far greater importance, as a rule, is the composition of the fuel and the completeness of its combustion. As an

---

[*] Boiler Efficiency, Capacity, and Smokelessness with Low-grade Fuels. Wm. H. Bryan. A paper read before the Engineers' Club of St. Louis, Oct. 21, 1896.

evaporation of 15.2 pounds of water from and at 212° is the theoretical limit of the calorific power of pure carbon, it is manifest that even the best coal cannot attain to this result in actual practice. In fact eleven pounds may be set as the practical limit under the very best conditions. As the quality of the fuel becomes lower and the character of the furnace inferior, the actual evaporation decreases until boiler-plants may be found in which only three or four pounds are being evaporated from and at 212° per pound of coal.

Although the primary purpose of a steam-boiler is to serve as a reservoir for water and steam, something more is required in the design than mere capacity and strength. In fact the boiler has a second and no less important office, namely, to utilize to the greatest possible extent the heat generated by the combustion of the fuel. Its actual efficiency is dependent upon the proportions and upon the arrangement of its surfaces for the transmission of the heat from the fire and gases to the water within. For this reason the simplicity of the plain cylinder-boiler is so far offset by its low efficiency that the more economical flue, tubular, and water-tube types have become almost universal substitutes.

Although the air required for supporting combustion is everywhere present, its passage to and through the fire in sufficient volume is absolutely dependent upon some means of draft-production. Upon the effectiveness of the device employed for this purpose depends the output, and to a certain extent the efficiency of the boiler to which it is applied.

Viewed as a whole, the efficiency of the fuel and of the boiler, including its furnace and means of draft-production, are practically interdependent. It is therefore difficult to determine in all cases the exact influence of any one of these various factors upon the ultimate efficiency.

Ultimate Efficiency of a Steam-boiler.—While other items may enter to affect the ultimate efficiency of a steam-boiler, it is fundamentally dependent upon three principal factors:

1. The primary cost of the entire plant and the fixed charges thereon.

2. The quantitative efficiency of the plant as a means of burning the fuel supplied and transferring the heat to the water evaporated.

3. The operating expenses, including the fuel.

In addition there are always distinct advantages or disadvantages incurred which, while of marked importance, can only be measured qualitatively in their relation to the superiority of any given method, arrangement, or appliance.

Primary Cost and Fixed Charges.—Value is not to be determined solely by first cost, but depends also upon durability and efficiency. The fixed charges are practically established by the rate of deterioration; for the interest, insurance, and taxes are substantially constant. In ordinary practice they will aggregate 6 to 7 per cent. But the amount to be charged to repairs and depreciation or sinking fund must be based upon the endurance of the given device. Evidently these items must be greater per year in the case of a device which becomes useless at the end of ten years, for instance, than in the case of one which proves serviceable for twenty years. Compared by twenty-year periods it is manifest that a higher first cost would be warranted in the second instance. In fact it would undoubtedly be necessary in order to secure the increased durability. The influence of the period of durability upon the fixed charges and permissible first cost is shown approximately by the accompanying Table No. 1, based upon a first cost of \$1000 for a device which would have to be entirely renewed at the end of five years.

In this table the interest on the investment has been taken at 5 per cent, the insurance and taxes at $1\frac{1}{2}$ per cent, and the repairs at $1\frac{1}{2}$ cent—a total of 8 per cent. Manifestly the annual repairs would be proportionately greatest for the period of least endurance; but inasmuch as the table is presented solely for the purpose of comparison, a fair average has been adopted.

TABLE No. 1.
COMPARATIVE FIRST COSTS AND FIXED CHARGES FOR DIFFERENT PERIODS OF ENDURANCE.

Period of Endurance in Years.	Annual Proportion for Fixed Cost of $1000.	Interest, Insurance, Taxes, and Repairs. 8% of First Cost.	Annual Payments to Sinking Fund earning 4½% Interest.	Total Annual Expense.	First Cost which would entail same Annual Expense as for Five-year Period.
5	$200.00	$80.00	$182.79	$462.79	$1000.00
10	100.00	80.00	81.38	261.38	1771.00
15	66.67	80.00	48.11	194.78	2376.00
20	50.00	80.00	31.87	161.87	2859.00
25	40.00	80.00	22.44	142.44	3242.00

With established types of boilers and accessories, the probable rate of deterioration can be reasonably well determined in advance, so that the question becomes one of business judgment as to the wisdom of a given expenditure to secure a given result.

But there is another item besides the durability to be taken into account in determining the actual value; namely, the efficiency of the specific boiler or appliance. Although this efficiency is primarily quantitative, it must be also measured commercially, as appears in what follows.

Quantitative Efficiency.—The proportion of the total calorific power of the fuel which is actually utilized for the generation of steam is dependent upon the means provided for the complete combustion of the fuel, with the minimum loss in the ash, and upon the arrangement and extent of the heating-surfaces of the boiler and of any supplementary heat-abstractor. The result is therefore intimately concerned with the design and consequently, to a great extent, with the first cost.

Quantitative efficiency is measured solely by the quantity of fuel supplied, as compared with the corresponding quantity of steam delivered, regardless of the cost. The results may be more or less affected by the perfection of the combustion, the means of supplying the air to the fuel, the amount of moisture in the air and the fuel, and by leakage of air and the radiation of heat through the settings.

In a properly designed plant the quantitative efficiency, as a rule, and within reasonable limits, increases with the first cost. But the attempt to secure higher efficiency may evidently be carried so far that the additional expenditure will nullify the resulting saving in fuel.

Thus, for instance, a feed-water economizer capable of cooling the escaping gases from 500° to 250° might prove to be an excellent investment, while one designed to abstract approximately 50 per cent more heat by cooling the gases to 125° would cost more in annual charges than it would save in the value of the fuel.

Operating Expenses.—By far the most important of the items of operating expense is the cost of the fuel. While this expense is closely related to the quantitative efficiency, it concerns it only in so far as the quantity of fuel is a measure of the cost of securing a given result. Thus a grade of coal evaporating 10.5 pounds of water from and at 212° per pound of coal, and costing \$3.50 per ton, may be less economical than one evaporating only 9 pounds and costing \$2.25 per ton. In the former instance the fuel cost of generating 1000 pounds of steam is 16.7 cents, while in the latter it is only 12.5 cents. In a 1000 H.P. plant this would represent an annual fuel saving of nearly \$4500.

As an offset to the economy thus indicated, the larger amount of coal will require more labor in handling and in removing the ashes, while special appliances or forms of construction may have to be introduced in order to secure the specified results.

Summary.—The preceding discussion, while brief, is sufficient to indicate that in steam-boiler practice many factors enter to determine the ultimate efficiency. In the succeeding chapters are presented the fundamental principles of economical operation, and the results obtained under a great variety of conditions. As far as possible the various influences for good or bad have been considered individually, so that they can be measured at their true value. It must be observed that to a great extent these influences are not cumulative. That is to say, decreased first cost of plant and increased fuel-efficiency seldom go hand in hand, and it is not possible, as a rule, to reduce the amount and cost of the fuel coincidently, or to maintain the financial saving secured by means of a heat-abstractor, and at the same time reduce the cost of the coal.

Intelligence, experience, and the judgment that is derived therefrom must be relied upon to reduce the losses that occur in modern steam-boiler practice, and to design, develop, and introduce the required means for the attainment of the desired results. Wherever these influences are brought to bear success is assured, if success is possible.

CHAPTER II.

WATER AND STEAM.

IN its bountiful presence in nature is to be found the principal reason for the employment of water in boiler and engine practice as the medium for absorbing heat during vaporization, and of subsequently transforming that heat into work when expansion occurs or of releasing it upon condensation.

Composition. — Pure water, whether solid, liquid, or gaseous, is a chemical combination of the two elements hydrogen and oxygen in the unalterable proportion of two parts, by volume, of the former to one part of the latter. A mere mechanical mixture of hydrogen and oxygen remains still a mixture until, through the influence of heat, electricity, or other special agents, the two chemically combine. If the union of the hydrogen and oxygen be effected in an apparatus so arranged that the water formed by the combination is kept at such a high temperature that it remains in the gaseous condition, it will be found that the two volumes of hydrogen and one volume of oxygen which were mixed together have become compacted into two volumes of steam, as the result of the chemical union.

Conversely, two volumes of water in its vaporized or gaseous form may, by various methods, be decomposed into its constituent elements; namely, two volumes of hydrogen and one volume of oxygen. Principal among the methods of dissociation is the application of heat. Consequently the presence of moisture in fuel may assume considerable importance in the ordinary process of combustion in connection with steam-boilers.

Water in its aeriform condition is to be considered as a vapor rather than as a gas, the term "gas" applying more properly to a body which, under ordinary temperature and pressure, continually remains in its aeriform state. In its vaporous condition water exists in the atmosphere in various proportions and under different conditions of atmospheric pressure, as indicated in Table No. 3. The slow process by which this production of aqueous vapor takes place at the free surface of a liquid is generally termed *evaporation*, while the more rapid production of the vapor throughout the mass is commonly designated as *boiling*. Under either method of vaporization the change is merely physical, the constituent elements remaining the same in character and proportion; namely, two parts of hydrogen and one part of oxygen, by volume.

Weight and Bulk.—Water is universally adopted as the standard by which the relative weights of all liquids and solids are determined, this relation being expressed by the term "specific gravity." The specific gravity of a body, therefore, indicates its weight as compared with that of an equal volume of pure water. Determinations of specific gravity are generally referred to the weight of one cubic foot of water at 62° F. At the more important temperatures the weights are as follows:

Weight of One Cubic Foot of Pure Water.

At 32° F. (freezing-point).................. 62.418 pounds
" 39.1° F. (maximum density)............. 62.425 "
" 62° F. (standard temperature)........... 62.355 "
" 212° F. (boiling-point under atmospheric pressure)......................... 59.640 "

For general purposes the weight of water is taken in round numbers as 62.5 pounds per cubic foot. The calculated weights at temperatures from 32° to 290° are given in Table No. 2.

TABLE No. 2.

EXPANSION AND DENSITY OF PURE WATER.

Temperature. Degrees Fahr.	Absolute Pressure of Vapor per Sq. In. Pounds.	Relative Volume. Water at 32° = 1.	Relative Density. Water at 32° = 1.	Density or Weight of 1 Cubic Foot. Pounds.	Temperature. Degrees Fahr.	Absolute Pressure of Vapor per Sq. In. Pounds.	Relative Volume. Water at 32° = 1.	Relative Density. Water at 32° = 1.	Density or Weight of 1 Cubic Foot. Pounds.
32°	0.089	1.00000	1.00000	62.418	135°	2.542	1.01539	0.98484	61.472
35	0.100	0.99993	1.00007	62.422	140	2.879	1.01690	0.98339	61.381
40	0.122	0.99989	1.00011	62.425	145	3.273	1.01839	0.98194	61.291
45	0.147	0.99993	1.00007	62.422	150	3.708	1.01989	0.98050	61.201
50	0.178	1.00015	0.99985	62.409	155	4.193	1.02164	0.97882	61.096
55	0.214	1.00038	0.99961	62.394	160	4.731	1.02340	0.97714	60.991
60	0.254	1.00074	0.99926	62.372	165	5.327	1.02589	0.97477	60.843
65	0.304	1.00119	0.99881	62.344	170	5.985	1.02690	0.97380	60.783
70	0.360	1.00160	0.99832	62.313	175	6.708	1.02906	0.97193	60.665
75	0.427	1.00239	0.99771	62.275	180	7.511	1.03100	0.97006	60.548
80	0.503	1.00299	0.99702	62.232	185	8.375	1.03300	0.96828	60.430
85	0.592	1.00379	0.99622	62.182	190	9.335	1.03500	0.96632	60.314
90	0.693	1.00459	0.99543	62.133	195	10.385	1.03700	0.96440	60.198
95	0.809	1.00554	0.99449	62.074	200	11.526	1.03889	0.96256	60.081
100	0.940	1.00639	0.99365	62.022	205	12.770	1.0414	0.9602	59.93
105	1.095	1.00739	0.99260	61.960	210	14.126	1.0434	0.9584	59.82
110	1.267	1.00889	0.99119	61.868	212	14.69	1.0444	0.9575	59.76
115	1.462	1.00989	0.99021	61.807	230	20.87	1.0529	0.9499	59.36
120	1.685	1.01139	0.98874	61.715	250	29.80	1.0628	0.9411	58.75
125	1.932	1.01239	0.98808	61.654	270	41.87	1.0727	0.9323	58.18
130	2.215	1.01390	0.98630	61.563	290	57.64	1.0838	0.9227	57.59

In bulk, water is usually measured by the gallon, the volume of which is 231 cubic inches (the British gallon contains 277.274 cubic inches), or 0.134 cubic feet. A gallon of water at 62°, therefore, weighs slightly over 8⅓ pounds, and 7.48 gallons equal one cubic foot.

When, by the application of heat, aqueous vapor is produced from, and in contact with, water in a closed vessel, it is usually denominated "steam," and under these conditions is always saturated. Saturated steam is of varying density and temperature, according to the pressure under which it is generated; but there exists an unalterable relation between density, elastic force, and temperature, such that if one of these properties remains constant the others so remain, while a change in one results in a change of the other two, and always in a fixed ratio.

The elastic force or pressure at low temperatures is shown in column 4 of Table No. 3, while that at higher temperatures, as expressed in pounds per square inch above vacuum, is presented in Table No. 7. In the last column of this table is given the weight of steam per cubic foot under the corresponding temperatures and pressures. The specific density of gaseous steam is 0.622, that of air being 1. That is to say, the weight of a cubic foot of gaseous steam is about five eighths of that of a cubic foot of air at same pressure and temperature.

Expansion by Heat.—Although water is practically noncompressible, even under the most extraordinary pressure, it readily expands by the mere application of heat, with the notable exception that between the temperature of melting ice at 32° and that of greatest density at 39.1° there is a gradual contraction in volume as heat is applied.

The rate of expansion is, however, variable, and above 212° but little is known regarding it experimentally. An application of the formula derived from experiments made below 212° gives at least approximate values. In Table No. 2 are embodied the results of calculation by Rankine's approximate formula, which gives substantially correct values at low and moderate temperatures. The results at high temperatures are, however, too large, as is evidenced by the fact that by the table the density at 212° is shown to be 59.76 pounds per cubic foot, while by actual measurement it has been found to be 59.64 pounds,—an error, however, of only 0.2 per cent in the value, which is fairly negligible except in refined calculations.

When saturated steam is superheated or surcharged with heat, as may only be done when it is separated from the water from which it was generated, it advances from the condition of saturation to that of gaseity. The gaseous state is only arrived at by considerably elevating the temperature, supposing the pressure remains the same. Obviously, the test of perfect gaseity must be the uniformity of the rate of expansion with the rise in temperature. Experiment has shown that, with the exception of a slight variation at a temperature just above that at which it was generated, steam, thus superheated, follows the law that controls the expansion of permanent gases; this law being that, the temperature remaining the same, the volume of a given quantity of gas is inversely proportional to the pressure which it sustains; or, conversely, that the pressure remaining the same, the volume will be proportional to the temperature. Therefore, the density or volume at any given temperature being known, it may be readily determined by proportion for any other temperature.

Absolute Zero.—The rate or coefficient of expansion of a perfect gas per degree, as determined by the most recent and refined experiments, is $0.00203 = \dfrac{1}{492.66}$ at the freezing-point, or 32° F. Or, in other words, for each rise in temperature of one degree, the gas of freezing temperature increases in volume $\dfrac{1}{492.66}$, and an increase of 492.66° would double the volume. If, then, this law holds good, reckoning upward from freezing-point, as has been conclusively proven by experiment, it is reasonable to suppose that it likewise holds good reckoning downward, and that for every degree of temperature withdrawn from the gas it is diminished $\dfrac{1}{492.66}$ of its volume at 32°. Carried to its limit, this would point to the fact that at 492.66° below freezing (or 460.66° below zero Fahrenheit) the contraction would be equal to the volume; that is, the volume would cease to exist. This is the so-called *absolute zero* of temperature, the starting-point for the scale of absolute temperatures by which the proportional expansion of gases is determined. Without appreciable error it may be

WATER AND STEAM. 13

TABLE No. 3.

WEIGHTS OF AIR, VAPOR OF WATER, AND SATURATED MIXTURES OF AIR AND VAPOR AT DIFFERENT TEMPERATURES, UNDER THE ORDINARY ATMOSPHERIC PRESSURE OF 29.921 INCHES OF MERCURY.

Temperature Degrees Fahrenheit.	Volume of Dry Air at Different Temperatures, the Volume at 32° being 1.000	Weight of a Cubic Foot of Dry Air at Different Temperatures, in Pounds.	Elastic Force of Vapor in Inches of Mercury, Regnault.	Mixtures of Air Saturated with Vapor.							Cubic Feet of Vapor from 1 lb. of Water at its own Pressure in Column 4.
				Elastic Force of the Air in the Mixture of Air and Vapor in Ins. of Mercury.	Weight of Cubic Foot of the Mixture of Air and Vapor.		Weight of the Vapor, in Pounds.	Total Weight of Mixture, in Pounds.	Weight of Vapor mixed with 1 lb. of Air, in Pounds.	Weight of Dry Air mixed with 1 lb. of Vapor, in Pounds.	
					Weight of the Air, in Pounds.						
1	2	3	4	5	6	7	8	9	10	11
0°	.935	.0864	.044	29.877	.0863	.000079	.086379	.00092	1092.4	3289
12	.960	.0842	.074	29.849	.0840	.000130	.084130	.00155	646.1	2252
22	.980	.0824	.118	29.803	.0821	.000202	.083202	.00245	406.4	
32	1.000	.0807	.181	29.740	.0802	.000304	.080504	.00379	263.81	
42	1.020	.0791	.267	29.654	.0784	.000440	.078840	.00561	178.18	
52	1.041	.0776	.388	29.533	.0766	.000627	.077227	.00819	122.17	1595
62	1.061	.0761	.556	29.365	.0747	.000881	.075581	.01179	84.70	1135
72	1.082	.0747	.785	29.136	.0727	.001221	.073921	.01680	59.54	819
82	1.102	.0733	1.092	28.829	.0706	.001667	.072267	.02361	42.35	600
92	1.122	.0720	1.501	28.420	.0684	.002250	.070717	.03289	30.40	444
102	1.143	.0707	2.036	27.885	.0659	.002997	.068897	.04547	21.98	334
112	1.163	.0694	2.731	27.190	.0631	.003946	.067046	.06253	15.99	253
122	1.184	.0682	3.621	26.300	.0599	.005142	.065042	.08584	11.65	194
132	1.204	.0671	4.752	25.169	.0564	.006639	.063039	.11777	8.49	151
142	1.224	.0660	6.165	23.756	.0524	.008473	.060873	.16170	6.18	118
152	1.245	.0649	7.930	21.991	.0477	.010716	.058416	.22465	4.45	93.3
162	1.265	.0638	10.099	19.822	.0423	.013415	.055375	.31173	3.15	74.5
172	1.285	.0628	12.758	17.163	.0360	.016682	.052682	.46338	2.16	59.2
182	1.306	.0618	15.960	13.961	.0288	.020536	.049336	.71300	1.402	48.6
192	1.326	.0609	19.828	10.093	.0205	.025142	.045642	1.22643	.815	39.8
202	1.347	.0600	24.450	5.471	.0109	.030545	.041445	2.80230	.357	32.7
212	1.367	.0591	29.921	0.000	.0000	.036820	.036820	Infinite	.000	27.1

expressed in round numbers as 461° below zero Fahrenheit, and will be so understood in calculations which follow.

In the case of a boiler, an increase in the temperature of the steam cannot occur without an equal increase in that of the water, which results in the generation of more steam and an increase in pressure; in fact, the relations are such that the temperature of the steam may always be determined from a knowledge of the pressure, and *vice versa*.

Specific Heat.—Bodies vary greatly in the capacity which they possess for absorbing heat under equal changes in temperature. The relation which thus exists between them is expressed by the "specific heat," which may be defined as the quantity of heat necessary to be imparted to a given body in order to raise its temperature one degree relatively to the quantity that is required to raise through one degree an equal weight of water at its point of greatest density at 39.1°. Thus, for instance, one pound of air at constant pressure may be raised through one degree by the expenditure of only 0.2375 of the heat necessary to raise one pound of water through one degree; or, what amounts to the same, the amount of heat expended to raise the temperature of one pound of water by one degree would heat $\dfrac{1}{0.2375} = 4.2105$ pounds of air through the same increment. As the specific heat of water is greater than that of any other known substance, the specific heat of all other substances must of necessity be expressed in decimals. Water does not absorb heat exactly in proportion to its increase in temperature; in other words, the specific heat of water varies with the temperature, as is rendered evident by Table No. 4.

The specific heat of saturated steam is 0.305 referred to water as a standard; that is, it requires only 0.305 as much heat to raise the temperature of a given weight of saturated steam through one degree as would be necessary to raise the same weight of water through the same increment. Properly, this value of 0.305 is the specific heat of the water and its saturated vapor combined. The specific heat of gaseous steam is 0.475.

TABLE No. 4.

SPECIFIC HEAT OF WATER.

Temperature. Degrees Fahrenheit.	Specific Heat at the Given Temperature. Freezing-point = 1.	Temperature. Degrees Fahrenheit.	Specific Heat at the Given Temperature. Freezing-point = 1.	Temperature. Degrees Fahrenheit.	Specific Heat at the Given Temperature. Freezing-point = 1.
32°	1.0000	176°	1.0089	320°	1.0294
50	1.0005	194	1.0109	338	1.0328
68	1.0012	212	1.0130	356	1.0364
86	1.0020	230	1.0153	374	1.0401
104	1.0030	248	1.0177	394	1.0440
122	1.0042	266	1.0204	410	1.0481
140	1.0056	284	1.0232	428	1.0524
158	1.0072	302	1.0262	446	1.0568

Unit of Heat.—The quantitative measure of heat is the thermal unit. The British thermal unit (as distinguished from the French thermal unit, or *calorie*) is that quantity of heat which is required to raise the temperature of one pound of pure water through one degree Fahr. at or near 39.1° Fahr., the temperature of maximum density of water. As employed in general practice, the term is usually abbreviated to "B. T. U." The *calorie* or French thermal unit is the quantity of heat absorbed by one gram or one kilogram of water (according to the unit adopted) when its temperature is increased one degree centigrade.

The relation existing between the temperature of water in degrees Fahrenheit and the number of thermal units contained therein, together with the increase in the number of thermal units for each increment of temperature of 5°, is indicated in Table No. 5.*

Mechanical Equivalent of Heat.—The mechanical unit of work is the "foot-pound," or the work required to raise one pound through a distance of one foot. The mechanical theory of heat regards heat as a mode of motion, and investi-

---

* From " Richards' Steam-engine Indicator." Charles T. Porter.

TABLE No. 5.

NUMBER OF THERMAL UNITS CONTAINED IN ONE POUND OF WATER.

Temperature. Degrees F.	Number of Thermal Units.	Increase.	Temperature. Degrees F.	Number of Thermal Units.	Increase.
35°	35.000		215°	215.939	5.065
40	40.001	5.001	220	221.007	5.068
45	45.002	5.001	225	226.078	5.071
50	50.003	5.001	230	231.153	5.075
55	55.006	5.003	235	236.232	5.079
60	60.009	5.003	240	241.313	5.081
65	65.014	5.005	245	246.398	5.085
70	70.020	5.006	250	251.487	5.089
75	75.027	5.007	255	256.579	5.092
80	80.036	5.009	260	261.674	5.095
85	85.045	5.009	265	266.774	5.100
90	90.055	5.010	270	271.878	5.104
95	95.067	5.012	275	276.985	5.107
100	100.080	5.013	280	282.095	5.110
105	105.095	5.015	285	287.210	5.115
110	110.110	5.015	290	292.329	5.119
115	115.129	5.019	295	297.452	5.123
120	120.149	5.020	300	302.580	5.128
125	125.169	5.020	305	307.712	5.132
130	130.192	5.023	310	312.848	5.136
135	135.217	5.025	315	317.988	5.140
140	140.245	5.028	320	323.134	5.146
145	145.175	5.030	325	328.284	5.150
150	150.305	5.030	330	333.438	5.154
155	155.339	5.034	335	338.596	5.158
160	160.374	5.035	340	343.759	5.163
165	165.413	5.039	345	348.927	5.168
170	170.453	5.040	350	354.101	5.174
175	175.497	5.044	355	359.280	5.179
180	180.542	5.045	360	364.464	5.184
185	185.591	5.049	365	369.653	5.189
190	190.643	5.052	370	374.846	5.193
195	195.697	5.054	375	380.044	5.198
200	200.753	5.056	380	385.247	5.203
205	205.813	5.060	385	390.456	5.209
210	210.874	5.061	390	395.672	5.216

gation has shown that there exists a definite relation between these two forms of energy, which is known as the "mechanical equivalent of heat." That is, if, as in the experiments of Joule, a certain known amount of mechanical energy is expended (as by the falling of a weight) to operate paddles in a vessel of water, the increase in temperature of the water, due to agitation by the paddles, will always be found to be proportional to the work done. This relation or proportion is universally expressed by the amount of work necessary to raise the temperature of one pound of water through one degree Fahrenheit, and the latest experimental determinations of Rowland show it to be practically 778 foot-pounds.

Latent and Sensible Heat.—When water is heated in the open atmosphere its temperature gradually increases until 212° is reached. Further application of heat has no effect in raising the temperature beyond this point, but the water boils and passes off as steam, the temperature of which will also be found to be 212°. Evidently, then, the heat applied to accomplish the vaporization cannot be measured by the thermometer. But experiment has shown that in the evaporation of the entire volume of water there thus disappears about five and a half times as much heat as is required to raise it from freezing- to boiling-point. On the other hand, it has been proven by experiment that upon the condensation of steam there is relinquished a large quantity of heat of which the thermometer gave no indication, and that this amount is exactly equal to that which disappeared in the process of vaporization. Because of the hidden character of this heat it is known as *latent heat*, and is measured in thermal units, while that indicated by the thermometer is designated as *sensible heat*. The sum of the heat-units in the water and in the steam generated therefrom is known as the *total heat*..

The total heat required in the formation of steam is expended in three ways.

1. In raising the temperature of the water to the boiling-point. This would be exactly measured by the sensible heat

if the specific heat of water was constant. The total heat of the water in heat-units varies slightly, however, from the temperature in degrees as indicated by the thermometer. This relation is clearly shown in Table No. 5.

2. In the work done in transforming the water into steam. This is distinctly an internal work, and consists in separating the water particles and establishing a repulsive action between them.

3. In the additional work of overcoming the incumbent pressure of the surrounding atmosphere so that enlargement of volume may take place. This work is entirely external.

An analysis of the heat and work expenditures in increasing the temperature of one pound of water from 32° to the boiling-point, and transforming it into steam of 212° temperature under atmospheric pressure, is displayed in Table No. 6,

TABLE No. 6.

HEAT AND WORK REQUIRED TO GENERATE ONE POUND OF SATURATED STEAM AT 212° FROM WATER AT 32°.

Distribution of Heat.	Units of Heat.	Mechanical Equivalent, in Foot-pounds.
The Sensible Heat. 1. To raise the temperature of water from 32° to 212°...	180.9	140740
The Latent Heat. 2. In the formation of steam	894.0	695532
3. In expansion against the atmospheric pressure	71.7	55783
Total heat and work..................	1146.6	892055

from which the relative amount of heat expended in each of the three ways is rendered evident.

Table No. 7,* in which are given the properties of saturated steam, taken in connection with Table No. 5, which gives the number of heat-units in the water for any given temperature, are of great convenience in all calculations relating to steam-boiler performance.

---

*From "Richards' Steam-engine Indicator." Charles T. Porter.

TABLE No. 7.

PROPERTIES OF SATURATED STEAM.

[From Charles T. Porter's treatise on *The Richards Steam-engine Indicator*.]

Pressure above zero.	Temperature.	Sensible Heat above zero Fahr.	Latent Heat.	Total Heat above zero Fahr.	Weight of One Cubic Foot.
Lbs. per sq. in.	Fahr. Deg.	B.T.U.	B.T.U.	B.T.U.	Lbs.
1	102.00	102.08	1042.96	1145.05	.0030
2	126.26	126.44	1026.01	1152.45	.0058
3	141.62	141.87	1015.25	1157.13	.0085
4	153.07	153.39	1007.22	1160.62	.0112
5	162.33	162.72	1000.72	1163.44	.0137
6	170.12	170.57	995.24	1165.82	.0163
7	176.91	177.42	990.47	1167.89	.0189
8	182.91	183.48	986.24	1169.72	.0214
9	188.31	188.94	982.43	1171.37	.0239
10	193.24	193.91	978.95	1172.87	.0264
11	197.76	198.49	975.76	1174.25	.0289
12	201.96	202.73	972.80	1175.53	.0313
13	205.88	206.70	970.02	1176.73	.0337
14	209.56	210.42	967.42	1177.85	.0362
15	213.02	213.93	964.97	1178.91	.0387
16	216.29	217.25	962.65	1179.90	.0413
17	219.41	220.40	960.45	1180.85	.0437
18	222.37	223.41	958.34	1181.76	.0462
19	225.20	226.28	956.34	1182.62	.0487
20	227.91	229.03	954.41	1183.45	.0511
21	230.51	231.67	952.57	1184.24	.0536
22	233.01	234.21	950.79	1185.00	.0561
23	235.43	236.67	949.07	1185.74	.0585
24	237.75	239.02	947.42	1186.45	.0610
25	240.00	241.31	945.82	1187.13	.0634
26	242.17	243.52	944.27	1187.80	.0658
27	244.28	245.67	942.77	1188.44	.0683
28	246.32	247.74	941.32	1189.06	.0707
29	248.31	249.76	939.90	1189.67	.0731
30	250.24	251.73	938.92	1190.26	.0755
31	252.12	253.64	937.18	1190.83	.0779
32	253.95	255.51	935.88	1191.39	.0803
33	255.73	257.32	934.60	1191.93	.0827
34	257.47	259.10	933.36	1192.46	.0851
35	259.17	260.83	932.15	1192.98	.0875
36	260.83	262.52	930.96	1193.49	.0899
37	262.45	264.18	929.80	1193.98	.0922
38	264.04	265.80	928.67	1194.47	.0946
39	265.59	267.38	927.56	1194.94	.0970
40	267.12	268.93	926.47	1195.41	.0994
41	268.61	270.46	925.40	1195.86	.1017
42	270.07	271.95	924.35	1196.31	.1041
43	271.50	273.41	923.33	1196.74	.1064
44	272.91	274.85	922.32	1197.17	.1088
45	274.29	276.26	921.33	1197.60	.1111
46	275.65	277.65	920.36	1198.01	.1134

STEAM-BOILER PRACTICE.

PROPERTIES OF SATURATED STEAM—*Continued.*

Pressure above zero.	Temperature.	Sensible Heat above zero Fahr.	Latent Heat.	Total Heat above zero Fahr.	Weight of One Cubic Foot.
Lbs. per sq. in.	Fahr. Deg.	B.T.U.	B.T.U.	B.T.U.	Lbs.
47	276.98	279.01	919.40	1198.42	.1158
48	278.29	280.35	918.46	1198.82	.1181
49	279.58	281.67	917.54	1199.21	.1204
50	280.85	282.96	916.63	1199.60	.1227
51	282.09	284.24	915.73	1199.98	.1251
52	283.82	285.49	914.85	1200.35	.1274
53	284.53	256.73	913.98	1200.72	.1297
54	285.72	287.95	913.13	1201.08	.1320
55	286.89	289.15	912.29	1201.44	.1343
56	288.05	290.33	911.46	1201.79	.1366
57	289.11	291.50	910.64	1202.14	.1388
58	290.31	292.65	909.83	1202.48	.1411
59	291.42	293.79	909.03	1202.82	.1434
60	292.52	294.91	908.24	1203.15	.1457
61	293.59	296.01	907.47	1203.48	.1479
62	294.66	297.10	906.70	1203.81	.1502
63	295.71	298.18	905.94	1204.13	.1524
64	296.75	299.24	905 20	1204.44	.1547
65	297.77	300.30	904.46	1204.76	.1569
66	298.78	301.33	903.73	1205.07	.1592
67	299.78	302.36	903.01	1205.37	.1614
68	300.77	303.87	902.29	1205.67	.1637
69	301.75	304.38	901.50	1205.97	.1659
70	302.71	305.37	900.80	1206.26	.1681
71	303.67	306.85	900.21	1206.56	.1703
72	304.61	307.32	899.52	1206.84	.1725
73	305.55	308.27	898.85	1207.13	.1748
74	306.47	309.22	898.18	1207.41	.1770
75	307.38	310.16	897.52	1207.69	.1792
76	308.29	311.09	896.87	1207.96	.1814
77	309.18	312.01	896.23	1208.24	.1836
78	310.06	312.92	895.59	1208.51	.1857
79	310.94	313.82	894.95	1208.77	.1879
80	311.81	314.71	894.33	1209.04	.1901
81	312.67	315.59	893.70	1209.30	.1923
82	313.52	316.46	893.09	1209.56	.1945
83	314.36	317.33	892.48	1209.82	.1967
84	315.19	318.19	891.83	1210.07	.1988
85	316.02	319.04	891.28	1210.32	.2010
86	316.83	319.88	890.69	1210.57	.2032
87	317.65	320.71	890.10	1210.82	.2053
88	318.45	321.54	889.52	1211.06	.2075
89	319.24	322.36	888.94	1211.31	.2097
90	320.03	323.17	888.37	1211.55	.2118
91	320.82	323.98	887.80	1211.79	.2132
92	321.59	324.78	887.24	1212.02	.2160
93	322.36	325.57	886.68	1212.26	.2182
94	323.12	326.35	886.13	1212.49	.2204
95	323.88	327.13	885.58	1212.72	.2224

WATER AND STEAM. 21

PROPERTIES OF SATURATED STEAM—*Continued.*

Pressure above zero.	Temperature.	Sensible Heat above zero Fahr.	Latent Heat.	Total Heat above zero Fahr.	Weight of One Cubic Foot.
Lbs.per sq. in.	Fahr. Deg.	B.T.U.	B.T.U.	B.T.U.	Lbs
96	324.63	327.90	885.04	1212.95	.2245
97	325.37	328.67	884.50	1213.18	.2266
98	326.11	329.43	883.97	1213.40	.2288
99	326.84	330.18	883.44	1213.62	.2309
100	327.57	330.93	882.91	1213.84	.2330
101	328.29	331.67	882.39	1214.06	.2351
102	329.00	332.41	881.87	1214.28	.2371
103	329.71	333.14	881.35	1214.50	.2392
104	330.41	333.86	880.84	1214.71	.2413
105	331.11	334.58	880.34	1214.92	.2434
106	331.80	335.30	879.84	1215.14	.2454
107	332.49	336.00	879.34	1215.35	.2475
108	333.17	336.71	878.84	1215.55	.2496
109	333.85	337.41	878.35	1215.76	.2516
110	334.52	338.10	877.86	1215.97	.2537
111	335.19	338.79	877.37	1216.17	.2558
112	335.85	339.47	876.89	1216.37	.2578
113	336.51	340.15	876.41	1216.57	.2599
114	337.16	340.83	875.94	1216.77	.2619
115	337.81	341.50	875.47	1216.97	.2640
116	338.45	342.16	875.00	1217.17	.2661
117	339.10	342.83	874.53	1217.36	.2681
118	339.73	343.48	874.07	1217.56	.2702
119	340.36	344.14	873.61	1217.75	.2722
120	340.99	344.78	873.15	1217.94	.2742
121	341.61	345.43	872.70	1218.13	.2762
122	342.23	346.07	872.25	1218.32	.2782
123	342.85	346.70	871.80	1218.51	.2802
124	343.46	347.34	871.35	1218.69	.2822
125	344.07	347.97	870.91	1218.88	.2842
126	344.67	348.59	870.47	1219.06	.2862
127	345.27	349.21	870.03	1219.25	.2882
128	345.87	349.83	869.59	1219.43	.2902
129	346.45	350.44	869.16	1219.61	.2922
130	347.05	351.05	868.73	1219.79	.2942
131	347.64	351.66	868.30	1219.97	.2961
132	348.22	352.26	867.88	1220.15	.2981
133	348.80	352.86	867.46	1220.33	.3001
134	349.38	353.46	867.03	1220.50	.3020
135	349.95	354.05	866.62	1220.67	.3040
136	350.52	354.64	866.20	1220.85	.3060
137	351.08	355.23	866.79	1221.02	.3079
138	351.75	355.81	865.38	1221.19	.3099
139	352.21	356.39	864.97	1221.36	.3118
140	352.76	356.96	864.56	1221.53	.3138
141	353.31	357.54	864.16	1221.70	.3158
142	353.86	358.11	863.76	1221.87	.3178
143	354.41	358.67	863.36	1222.03	.3199
144	354.96	359.24	862.96	1222.20	.3219

PROPERTIES OF SATURATED STEAM—*Continued.*

Pressure above zero.	Temperature.	Sensible Heat above zero Fahr.	Latent Heat.	Total Heat above zero Fahr.	Weight of One Cubic Foot.
Lbs. per sq. in.	Fahr. Deg.	B.T.U.	B.T.U.	B.T.U.	Lbs.
145	355.50	359.80	862.56	1222.36	.8239
146	356.03	360.85	862.17	1222.53	.3259
147	356.57	360.91	861.78	1222.69	.3279
148	357.10	361.46	861.39	1222.85	.3290
149	357.63	362.01	861.00	1223.01	.3319
150	358.16	362.55	860.62	1223.18	.3340
151	358.68	363.10	860.23	1223.33	.3358
152	359.20	363.64	859.85	1223.40	.3376
153	359.72	364.17	859.47	1223.65	.3394
154	360.23	364.71	859.10	1223.81	.3412
155	360.74	365.24	858.73	1223.97	.3430
156	361.26	365.77	858.35	1224.12	.3448
157	361.76	366.30	857.98	1224.28	.3466
158	362.27	366.82	857.61	1224.43	.3484
159	362.77	367.34	857.24	1224.58	.3502
160	363.27	367.86	856.87	1224.74	.3520
161	363.77	368.38	856.50	1224.89	.3539
162	364.27	368.89	856.14	1225.04	.3558
163	364.76	369.41	855.78	1225.19	.3577
164	365.25	369.92	855.42	1225.34	.3596
165	365.74	370.42	855.06	1225.49	.3614
166	366.23	370.93	854.70	1225.64	.3633
167	366.71	371.43	854.35	1225.78	.3652
168	367.19	371.93	853.99	1225.93	.3671
169	367.68	372.43	853.64	1226.08	.3690
170	368.15	372.93	853.29	1226.22	.3709
171	368.63	373.42	852.94	1226.37	.3727
172	369.10	373.91	852.59	1226.51	.3745
173	369.57	374.40	852.25	1226.66	.3763
174	370.04	374.89	851.90	1226.80	.3781
175	370.51	375.38	851.56	1226.94	.3799
176	370.97	375.86	851.22	1227.08	.3817
177	371.44	376.34	850.88	1227.23	.3835
178	371.90	376.82	850.54	1227.37	.3853
179	372.36	377.30	850.20	1227.51	.3871
180	372.82	377.78	849.86	1227.65	.3889
181	373.27	378.25	849.53	1227.78	.3907
182	373.73	378.72	849.20	1227.92	.3925
183	374.18	379.19	848.86	1228.06	.3944
184	374.63	379.66	848.53	1228.20	.3962
185	375.08	380.13	848.20	1228.33	.3980
186	375.52	380.59	847.88	1228.47	.3999
187	375.97	381.05	847.55	1228.61	.4017
188	376.41	381.51	847.22	1228.74	.4035
189	376.85	381.97	846.90	1228.87	.4053
190	377.29	382.42	846.58	1229.01	.4072
191	377.72	382.88	846.26	1229.14	.4089
192	378.16	383.33	845.94	1229.27	.4107
193	378.59	383.78	845.62	1229.41	.4125

Properties of Saturated Steam—*Continued.*

Pressure above zero, Lbs. per sq. in.	Temperature, Fahr. Deg.	Sensible Heat above zero Fahr, B.T.U.	Latent Heat, B.T.U.	Total Heat above zero Fahr, B.T.U.	Weight of One Cubic Foot, Lbs.
194	379.02	384.23	845.30	1229.54	.4143
195	379.45	384.67	844.99	1229.67	.4160
196	379.97	385.12	844.68	1229.80	.4178
197	380.30	385.56	844.36	1229.93	.4196
198	380.72	386.00	844.05	1230.00	.4214
199	381.15	386.44	843.74	1230.19	.4231
200	381.57	386.88	843.43	1230.31	.4249
201	381.99	387.32	843.12	1230.44	.4266
202	382.41	387.76	842.81	1230.57	.4283
203	382.82	388.19	842.50	1230.70	.4300
204	383.24	388.62	842.20	1230.82	.4318
205	383.65	389.05	841.89	1230.95	.4335
206	384.06	389.48	841.59	1231.07	.4352
207	384.47	389.91	841.29	1231.20	.4369
208	384.88	390.33	840.99	1231.32	.4386
209	385.28	390.75	840.69	1231.45	.4403
210	385.67	391.17	840.39	1231.57	.4421

CHAPTER III.

COMBUSTION.

Definition.—The phenomenon of combustion may result from any chemical action in which the energy is sufficient to heat the body to the point of luminosity. But in its more restricted sense, as regards steam-boiler practice, combustion may be defined as a process of rapid oxidation, caused by the chemical union of the oxygen in the air with any material which is capable of oxidation. A combustible is therefore to be considered as any substance capable of combining rapidly with oxygen so as to produce light and heat, while oxygen is to be classed as a supporter of combustion.

Carbon.—Of all combustibles carbon is the most widely distributed, and readily obtained in nature. Because of its abundance as a constituent of coal, wood, peat, mineral oil, and natural gas, these substances are almost exclusively adopted as fuels. To these may be added coke, charcoal, and fuel-gas, which are produced by special processes from these natural substances. Carbon itself is an infusible, non-volatile solid, of which three distinct modifications occur; viz., (1) diamond, (2) plumbago or graphite, (3) charcoal or lamp-black. Among natural fuels—that is, those not prepared by artificial means—anthracite coal most nearly approaches to the condition of pure carbon, and is to be classed between graphite and charcoal.

Oxygen.—Although oxygen, the universal supporter of combustion, as here defined, is the most abundant of all natural substances, it never exists by itself in nature, but always in association with some other substance. Thus it is that as a constituent of atmospheric air it is associated with the inert gas, nitrogen; the relative proportion, of the two gases in pure air remaining practically constant under all conditions.

As determined by recent investigation, pure air is composed, by volume, of—

    Oxygen. .... 0.2096 parts
    Nitrogen............... 0.7904 "
                             1.0000 parts

and by weight of—

    Oxygen................ 0.2317 parts
    Nitrogen............... 0.7683 "
                             1.0000 parts

In nature, however, this proportionate composition of pure air is slightly affected by the presence of aqueous vapor, carbonic acid, and other impurities. Unless extreme accuracy is desired, it is usually convenient to consider the atmosphere as composed of one volume of oxygen and four volumes of nitrogen. Air is, however, but a mechanical mixture of the two gases, and the oxygen is therefore free, without chemical dissociation, to leave the nitrogen and unite with other substances, which it does with great avidity under favorable circumstances. In its independent state oxygen is colorless, tasteless, and slightly heavier than air, in the proportion of 1 to 1.1056.

The Atomic Theory.—A clear understanding of the atomic theory is necessary to a full comprehension of the principles of combustion. This theory, which has been developed through years of investigation, is now universally accepted as the explanation of all chemical phenomena.

By experiment it has been demonstrated that all chemical combinations between elementary substances are made in definite and invariable proportions. For instance, if hydrogen and oxygen be mixed and caused to form water, as already described, it will be found that the entire amount of these two gases will be utilized and enter into combination only when they originally existed in the exact proportion of two volumes of hydrogen to one volume of oxygen. Observation also proves that if this water is maintained in its gaseous condition it will occupy only the space of two volumes, although for its production a total of three volumes was supplied.

Two volumes of hydrogen and one and a half volumes of oxygen cannot be made to chemically combine to form the compound water, for the hydrogen will unite with only its proportional quantity, leaving the extra half-volume of oxygen unassociated. No matter how large or how small these volumes may be, the same relation holds. It is, therefore, reasonable to suppose that if the smallest conceivable particle of oxygen be brought into union with two of the smallest conceivable particles of hydrogen, there will be the same result and a minute particle of water will be formed.

These minute particles, the smallest in which elementary substances may be conceived to enter into combination with each other, are called atoms, while the individual particles resulting from their union are known as molecules. From the above reasoning it would appear probable that equal volumes of the elementary gases, at least, contain the same number of atoms, and therefore that the atoms are of equal size. Although attempts have been made to calculate the probable dimensions of these atoms, we have no direct knowledge as to their size.

Chemists have adopted as designating symbols for the various elements the initials of their names, followed, when necessary for distinction, by a succeeding letter. Thus hydrogen is designated by H and oxygen by O. The compound water, formed by the chemical union of two atoms of hydrogen and one atom of oxygen, can therefore be simply represented by H_2O, the suffix 2 being employed to indicate the presence by volume of twice as much hydrogen as oxygen.

Upon the assumption that the atoms are of equal size, the determination of the relative weights of equal volumes of these two gases, under the same pressure and temperature, is equivalent to determining the relative weights of the atoms themselves; that is, their *atomic weights*. The weight of hydrogen, which is the lightest of all known substances, is taken as unity, the relative weight of oxygen being 16. That is, a given volume of oxygen weighs 16 times as much as an equal volume of hydrogen.

The symbol H_2O, therefore, reveals still another fact as to the composition of water; namely, that 2 atoms of hydrogen, weighing relatively $2 \times 1 = 2$, are combined with 1 atom of oxygen, weighing 16. In other words, that by weight water is composed of 2 parts of hydrogen and 16 parts of oxygen; or more simply, that the ratio of the hydrogen to the oxygen is as 1 to 8.

But in this process of combination it has already been shown that the two volumes of hydrogen and one volume of oxygen unite to form only two volumes of water in its gaseous state, which two volumes represent the space originally occupied by the hydrogen. Hence it is evident that the compound now weighing 18 occupies the same space as an amount of hydrogen weighing 2, and that its relative density, the density of hydrogen being taken as unity, is $\frac{18}{2} = 9$. In other words, gaseous water of given temperature and pressure weighs nine times as much as an equal volume of hydrogen under the same conditions.

The common elementary substances entering into the composition of fuel, with their symbols and atomic weights in round numbers, are given in Table No. 8.

Union of Carbon and Oxygen.—Many elements enter into chemical combination with each other in more than one proportion. This is true of carbon and oxygen. If a piece of carbon, heated to incandescence, be placed in a sufficient volume of oxygen or air, each atom of the carbon will unite with two atoms of the oxygen to form a compound known as

TABLE No. 8.

SYMBOLS AND ATOMIC WEIGHTS OF ELEMENTARY SUBSTANCES CONCERNED IN COMBUSTION.

Name.	Symbol.	Atomic Weight.
Hydrogen	H.	1
Carbon	C.	12
Nitrogen	N.	14
Oxygen	O.	16
Sulphur	S.	32

carbonic acid, or carbonic dioxide, the symbol of which is CO_2; the process being indicated by the formula $C + 2O = CO_2$. No matter how plenteous the oxygen, it cannot be made to enter into combination with the carbon in a proportion greater than two atoms of oxygen to one atom of carbon.

This gas is, therefore, evidently a product of complete combustion, there having been a full supply of oxygen. As is shown by Table No. 8, the single atom of carbon weighs 12 relatively to each atom of oxygen, which weighs 16; that is, the compound consists by weight of 12 parts of carbon and $2 \times 16 = 32$ parts of oxygen.

Carbonic acid gas is transparent and colorless, about one and a half times heavier than air, and of a slightly acid taste and smell. It is incombustible, being already the product of complete combustion, and, although not directly poisonous, is neither a supporter of animal life nor of combustion.

If, in turn, this gas, without the accompaniment of sufficient oxygen, be brought into contact with incandescent carbon, it will be deprived of one half its oxygen, each atom of oxygen thus released uniting with an atom of carbon to form a new compound known as carbonic oxide, with the symbol CO. The process of combination may be symbolically expressed thus: $CO_2 + C = 2CO$, showing that not only is the new compound formed by union of carbon with the released oxygen, but that the carbonic acid thus deprived of its oxygen is thereby also reduced to carbonic oxide. The relative weights of the elements of this compound are, evidently, carbon = 12 and oxygen = 16. This gas is slightly lighter than air, transparent, colorless, and practically odorless, and is destructive to animal life, being in fact a direct poison. It is not a supporter of combustion, but, being the product of imperfect combustion, is itself a combustible and may be readily burned in the air. Such being the case, we should expect that the process of burning—which has already been defined as the rapid chemical union of a combustible with oxygen—would result in an accession of oxygen to the carbonic oxide. Experiment will prove this to be true, the product being carbonic acid, the same compound already shown to be the result of complete combustion. Symbolically, the process is expressed by $CO + O = CO_2$. In tabular form, the general properties of carbonic oxide and carbonic acid are shown in Table No. 9.

TABLE No. 9.

PROPERTIES OF CARBONIC OXIDE AND CARBONIC ACID.

Name.	Symbol.	Composition.					
		By Weight.			Percentage.		
		Carbon.	Oxygen.	Total.	Carbon.	Oxygen.	Total.
Carbonic oxide..	CO	12	16	28	42.86	57.14	100
Carbonic acid...	CO_2	12	32	44	27.27	72.73	100

Combustion of Fuel.—The two elements contributing most largely to the economic value of any fuel, as measured by its heating power, are carbon and hydrogen. These elements exist in fuels either combined, or, upon the application of heat, associate themselves in a series of complex compounds known as hydrocarbons, the simplest of the list of some fifty being marsh-gas, represented by the symbol CH_4. Such portion of the carbon or hydrogen as does not thus enter into combination, and for which there exists in the entire substance no further material for combination, is designated as fixed.

Besides these primary elements, fuels usually contain small amounts of oxygen, nitrogen, and sulphur, together with a certain percentage of incombustible matter which remains as ash after the process of combustion is complete. The phenomena attendant upon the combustion of ordinary fuels are, therefore, much more complex than those resulting from the combustion of carbon alone. Although it must be evident that fuels of the same general character vary considerably in the proportions of their constituents, their relative average elementary composition by weight is substantially as shown in Table No. 10. The results there given are those deter-

TABLE No. 10.

COMPOSITION OF FUELS.

Description.	Carbon.	Hydrogen.	Oxygen.	Nitrogen.	Sulphur.	Ash.
ANTHRACITES:						
France..................	90.9	1.47	1.53	1.00	0.80	4.3
Wales...................	91.7	3.78	1.30	1.00	0.72	1.5
Rhode Island..........	85.0	3.71	2.39	1.00	0.90	7.0
Pennsylvania..........	78.6	2.5	1.7	0.8	0.4	14.8
SEMI-BITUMINOUS:						
Maryland	80.0	5.0	2.7	1.1	1.2	8.3
Wales..................	88.3	4.7	0.6	1.4	1.8	3.2
BITUMINOUS:						
Pennsylvania.........	75.5	4.93	12.35	1.12	1.10	5.0
Indiana	69.7	5.10	19.17	1.23	1.30	3.5
Illinois................	61.4	4.87	35.42	1.41	1.20	5.7
Virginia...............	57.0	4.96	26.44	1.70	1.50	8.4
Alabama...............	53.2	4.81	32.37	1.62	1.30	6.7
Kentucky..............	49.1	4.95	41.13	1.70	1.40	7.2
Cape Breton..........	67.2	4.26	20.16	1.07	1.21	6.1
Vancouver's Island...	66.9	5.32	8.76	1.02	2.20	15.8
Lancashire gas-coal...	80.1	5.5	8.1	2.1	1.5	2.7
Boghead cannel......	63.1	8.9	7.0	0.2	1.0	19.8
LIGNITES:						
California brown coal.	49.7	3.78	30.19	1.0	1.53	13.8
Australian brown coal.	73.2	4.71	12.35	1.11	0.63	8.0
PETROLEUMS:						
Pennsylvania, crude..	84.9	13.7	1.4	....	....	......
Caucasian, light......	86.3	13.6	0.1	....	....	......
Caucasian, heavy.....	86.6	12.3	1.1	....	....	......
Refuse.................	87.1	11.7	1.2	....	....	......

mined by ultimate analysis. For the general purposes of comparison of fuels, the method of proximate analysis, whereby only the relative percentages of carbon, volatile matter, ash, and moisture are ascertained, is sufficiently refined.

Obviously, owing to the conditions under which combustion takes place, it is impossible to determine in detail the exact order of the process. It is certain, however, that the final results of perfect combustion of ordinary fuel should be carbonic acid gas (CO_2), water (H_2O), nitrogen (N), and possibly a little sulphurous acid (SO_2). The process may be outlined as follows: If, for instance, coal of bituminous character be thrown upon a glowing fire, the heat first volatilizes and frees the hydrocarbons, at comparatively low temperature. These inflammable gases are thereupon immediately ignited, and by the heat thus produced assist in bringing the remainder of the coal to a state of incandescence. The burning of the hydrocarbons is indicative of their union with oxygen, whereby these compounds are broken up and new combinations of a simpler character are formed. The three elements thus presented for combination are carbon, hydrogen, and oxygen. If the supply of oxygen is sufficient, the carbon leaves the hydrogen with which it has been associated, and unites with the oxygen to form carbonic acid, an evidence that combustion is complete. The liberated hydrogen also unites with oxygen, if a sufficiency is present, and forms water, which, being at a high temperature, is maintained in its gaseous condition. If, upon dissociation, a portion of the carbon which is liberated in incandescent particles does not immediately meet with its complement of oxygen, it is liable to become cooled to such an extent by the surrounding gases that, when it reaches an abundance of oxygen, its temperature will be too low to permit of chemical union. It will therefore pass off as unconsumed and visible carbon, in the form of smoke.

By the time the coal has become incandescent all of the

hydrocarbons will have been expelled, and the carbon will be in a condition to enter into combination with the oxygen of the air, or of any surrounding carbonic acid. If oxygen is present in excess, the product will be carbonic acid; but if carbonic acid be brought into contact with the glowing coal, carbonic oxide will be the result. This in turn will burn to carbonic acid, if only supplied with sufficient air.

In consideration of the number and great variety of interstices existing between the lumps of coal, and of the various stages of combustion to which different portions of the fuel have attained, it is evident that in their passage through the fire many changes must take place in the composition of the gases. Association and dissociation must follow in rapid succession; at one instant an atom of carbon may be combined with two atoms of oxygen to form carbonic acid, while in the next it may have lost one of its atoms of oxygen and have been reduced to carbonic oxide, which in turn may come in contact with a sufficiency of oxygen to again form carbonic acid. In just which combination the carbon and oxygen shall leave the fire and pass to the chimney must depend upon the temperature of the gases and the proportion of oxygen at hand.

Air Required for Combustion.—The definite proportions in which oxygen unites with hydrogen and carbon to form water and carbonic acid—the results of perfect combustion—have already been shown. Expressed in pounds, one pound of hydrogen requires eight pounds of oxygen for its complete combustion; for the atomic weights are respectively $H = 1$ and $O = 16$, and the composition of water, which is the product of perfect combustion of hydrogen and oxygen, is symbolically indicated by H_2O. Hence, substituting atomic weights for symbols, $H_2O = 2 + 16$, and $H_2 : O :: 2 : 16 = H_2 : O :: 1 : 8$.

For the complete combustion of one pound of carbon there are required $2\frac{2}{3}$ pounds of oxygen, for, by atomic weights,

COMBUSTION. 33

$C = 12$ and $O = 16$; hence $CO_2 = 12 + (2 \times 16)$ and $C : O_2 :: 12 : 32 = C : O_2 :: 1 : 2\tfrac{2}{3}$.

Air consists, by weight, of 0.2317 parts oxygen; therefore the amount of air required for the combustion of one pound of carbon must be the amount that would contain $2\tfrac{2}{3}$ pounds of oxygen; that is, $2\tfrac{2}{3} \div 0.2317 = 11.51$ pounds.

In Table No. 11 are presented the principal data regarding oxygen, nitrogen, and the elements of the common combustibles, together with the amount of oxygen and air required for each as calculated in the manner just described.

TABLE No. 11.

COMBUSTION DATA.

Combustible.				Product of Combustion.				Required per Pound of Combustible.		
Name.	Symbol.	Atomic or Molecular Weight.	Density or Weight of 1 Cubic Foot. lbs.	Name.	Symbol.	Atomic or Molecular Weight.	Density or Weight of 1 Cubic Foot. lbs.	Oxygen. lbs.	Air. lbs.	Air. Cubic Feet at 62°.
				Oxygen.........	O	16	0.08928			
				Nitrogen......	N	14	0.07837			
Hydrogen	H	1	0.00559	Water	H_2O	18	....	8.000	34.52	454
Carbon.........	C	12	....	Carbonic oxide.	CO	28	0.07806	1.333	5.76	76
Carbon.........	C	12	....	Carbonic acid..	CO_2	44	0.12341	2.667	11.51	151
Carbonic oxide.	CO	28	0.07806	Carbonic acid..	CO_2	44	0.12341	0.571	2.45	32
Marsh-gas.. ...	CH_4	16	0.04464	{ Water.........	H_2O	18	....	} 4.000	17.26	227
				{ Carbonic acid	CO_2	44	0.12341			
Olefiant gas....	C_2H_4	28	0.07809	{ Water.........	H_2O	18	....	} 3.428	14.80	195
				{ Carbonic acid	CO_2	44	0.12341			
Sulphur.........	S	32	....	Sulphurous acid	SO_2	64	0.17860	1.000	4.32	57

As already shown in Table No. 10, oxygen enters to a certain extent into the original composition of all fuels. In the process of combustion, this oxygen unites with its equivalent of hydrogen, which is thus rendered inert, so far as combination with extraneous oxygen is concerned. In calculation, therefore, this quantity of hydrogen is disregarded, and there are left to be considered only the remaining carbon and hydrogen.

34   STEAM-BOILER PRACTICE.

The method of calculation of the amount of air necessary for the combustion of ordinary coal can best be explained by means of an example based upon the known composition of a certain fuel, as, for instance, that of the Maryland semi-bituminous coal, given in Table No. 10. For simplicity these proportionate figures are here given in pounds instead of in per cent:

Carbon .... 80.0 pounds
Hydrogen .... 5.0   "
Oxygen .... 2.7   "
Nitrogen .... 1.1   "
Sulphur .... 1.2   "
Ash. .... 8.3   "

The nitrogen is inert, the sulphur, because of the small amount in which it is present, may be disregarded, and the ash, being incombustible, has no appreciable effect upon the result so far as the chemical requirements are concerned.

We may therefore estimate as follows for the amount of oxygen required: The 2.7 pounds of oxygen will render inert $\frac{2.7}{8} = 0.3375$ pounds of hydrogen, for it will directly combine with that amount. The constituents to be considered thus become—

Carbon 80   pounds
Hydrogen, 5.0 — 0.3375 = ...   4.6625   "
                                    ───────
                                    84.6625   "

and their requirements in the way of oxygen will be—

Carbon, 80 × 2⅔ = ... 213.33 pounds
Hydrogen, 4.6625 × 8 = ...   37.30   "
                                  ──────
                                  250.63   "

The weight of air containing the above quantity of oxygen is $\frac{250.63}{0.2317} = 1081.7$ pounds. The quantity required per pound of combustible will therefore be $\dfrac{1081.7}{84.6625} = 12.77$ pounds. As these constituents are part of a quantity of coal weighing 100 pounds when its elementary moisture is included, the amount of air necessary per pound of this particular coal is $\dfrac{1081.7}{100} = 10.82$ pounds.

For approximate calculation of the weight of air required for combustion the following formula may be used:

$$\text{Weight of air} = 12C + 36\left(H - \frac{O}{8}\right).$$

In this equation the weights of carbon, hydrogen, and oxygen are represented respectively by their symbols, C, H, and O, the amount of hydrogen rendered inert by the oxygen in the fuel is allowed for, and the proportion of oxygen and nitrogen in the atmosphere is taken as one to four.

As is evident by what follows, it is unnecessary, for practical purposes, to compute with great exactness the weight of air necessary for the combustion of fuel; for the excess of air which is usually supplied, together with the variableness in the composition of fuel, render all ordinary calculations somewhat approximate. It is therefore the common practice to estimate the approximate amount required, for either coke or coal (and with the same approximation for pure carbon), at 12 pounds per pound of fuel.

The air required per pound of fuel for various fuels of typical composition on a basis of 12 pounds per pound of carbon is shown in Table No. 12.*

Air for Dilution.—The preceding calculations of air-supply are based upon the assumption that each individual atom of oxygen in the air comes in contact and unites with its proportion of hydrogen or carbon in the fuel. When it is considered that this oxygen is intimately united with about four times its

---

* "A Manual of the Steam-engine." W. J. M. Rankine.

TABLE No. 12.

AIR REQUIRED FOR COMBUSTION OF FUELS.

Fuel.	Weight of Given Constituent in 1 Pound of Fuel.			Air Required per Pound of Dry Fuel. Pounds.
	Carbon.	Hydrogen.	Oxygen.	
Charcoal—From wood..	0.93			11.16
From peat...	0.80			9.6
Coke, good	0.94			11.28
Coal—Anthracite.......	0.915	0.035	0.026	12.13
Dry bituminous..	0.87	0.05	0.04	12.06
Coking...........	0.85	0.05	0.06	11.73
Coking...........	0.75	0.05	0.05	10.58
Cannel...........	0.84	0.06	0.08	11.88
Dry, long-flaming	0.77	0.05	0.15	10.32
Lignite...........	0.70	0.05	0.20	9.30
Peat, dry.................	0.58	0.06	0.31	7.68
Wood, dry...............	0.50			6.00
Mineral oil...............	0.85			15.65

volume of nitrogen, whereby it is to a certain extent separated from the fuel, and further that the variety in the arrangement of the fuel and the passages through it affect any attempt at equal distribution of the air, it must be evident that the above assumption cannot ordinarily be maintained in practice. It therefore usually becomes necessary in practice to furnish sufficient air in excess of the calculated amount to insure complete combustion in all parts of the furnace.

It has already been shown that one pound of pure carbon demands for its complete combustion $2\frac{2}{3}$ pounds of oxygen, which in turn is a constituent part of 11.51 pounds of air. If no air is supplied in excess of that chemically required, the final products of combustion will be $3\frac{2}{3}$ pounds of carbonic acid (consisting of one pound of carbon chemically united with $2\frac{2}{3}$ pounds of oxygen) and 8.84 pounds of nitrogen (which, in mechanical combination with $2\frac{2}{3}$ pounds of oxygen, constituted the original 11.51 pounds of air). Had 100 per cent excess of air been supplied its weight would have been $2 \times 11.51 =$ 23.02 pounds, of which $2\frac{2}{3}$ pounds of oxygen would have united with the pound of carbon to form, as before, $3\frac{2}{3}$ pounds of carbonic acid. The remaining products of combustion would be $11.51 \times 2 - 2.67 = 20.35$ pounds of nitrogen and $2\frac{2}{8}$ pounds of oxygen. In a similar manner it may be shown that with 200 per cent excess of air the carbonic acid would remain the same, but there would also be 29.19 pounds of nitrogen and 5.34 pounds of oxygen.

The relative proportions of these products of combustion of one pound of pure carbon with different amounts of air in excess are graphically indicated in the accompanying diagram, Fig. 1. Here the volume of CO_2 has been taken as unity, and

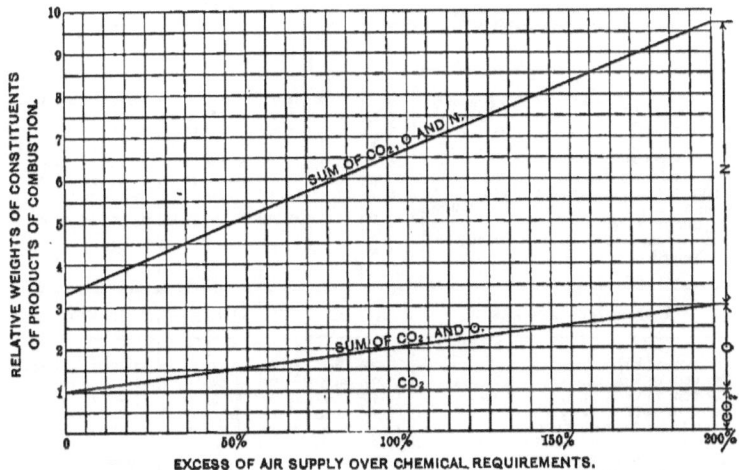

FIG. 1. RELATIVE COMPOSITION OF PRODUCTS OF PERFECT COMBUSTION OF PURE CARBON WITH VARIOUS WEIGHTS OF AIR IN EXCESS.

it appears, for instance, that with no excess of air the total weight of all the products is $\dfrac{1 + 11.51}{1 + 2.67} = 3.41$ times that of the carbonic acid alone.

Evidently, the amount of air supplied for dilution must vary greatly in different cases. This is clearly shown by the results of numerous careful tests of different boilers by Messrs. Donkin and Kennedy. In each case the volume of air supplied was determined by chemical analysis of the furnace-gases, the results of which, together with the deductions relating to the amount of air supplied, are presented in Table

No. 13. It will be noted that the dry air supplied, per pound of coal, ranges from 16.1 to 40.7 pounds, and that the corresponding ratio of air used to air theoretically required ranges from 1.56 to 4.28; that is, from 56 per cent to 328 per cent in excess. The composition of the gases is given by weight.

TABLE No. 13.
ANALYSIS AND CALCULATIONS RELATING TO FURNACE-GASES AND AIR-SUPPLY.

Analysis and Calculations.	Number of Test.							
	I	II	III	IV	V	VI	VII	VIII
Per cent of CO_2	15.15	13.00	18.21	11.71	7.90	10.44	11.00	14.95
Per cent of CO	2.59	0.00	0.24	0.00	0.00	0.28	0.10	0.28
Per cent of O	6.46	11.15	7.55	13.13	17.00	13.37	13.20	8.31
Per cent of N	75.80	75.85	74.00	75.16	75.10	75.90	75.70	76.46
Pounds dry air per pound of C	18.50	27.60	19.20	30.70	46.00	33.10	32.30	23.30
Pounds dry air per pound of coal	16.40	24.40	17.00	27.20	40.70	29.30	28.60	20.60
Ditto, per pound pure dry coal	16.90	25.20	17.50	28.00	42.20	30.30	29.60	21.20
Pounds dry furnace-gases per pound pure dry coal	17.50	25.80	18.10	28.60	42.80	30.90	30.20	21.80
Ratio of air used to air theoretically required	1.58	2.40	1.63	2.61	4.28	2.82	2.76	1.98

Analysis and Calculations.	IX	XI	XII	XIII	XIV	XVII	XIX	XX
Per cent of CO_2	8.60	16.50	15.10	17.94	14.90	11.10	11.53	18.88
Per cent of CO	0.00	0.21	0.00	1.02	0.00	0.00	0.00	0.34
Per cent of O	14.40	7.76	8.80	6.53	6.60	13.10	13.03	5.85
Per cent of N	77.00	75.53	76.10	74.51	78.50	75.80	75.44	74.93
Pounds dry air per pound of C	42.10	21.20	23.70	18.20	24.00	32.70	31 20	18.30
Pounds dry air per pound of coal	37.30	18.80	21.00	16.10	21.30	29.00	27.6c	16.20
Ditto, per pound pure dry coal	38.60	19.40	21.70	16.70	22.00	30.00	28.60	16.80
Pounds dry furnace-gases per pound pure dry coal	39.20	20.00	22.30	17.30	22.60	30.60	29.20	17.40
Ratio of air used to air theoretically required	3.60	1.81	2.02	1.56	2.05	2.80	2.67	1.56

Accepting 12 pounds of air, per pound of fuel, as in round numbers the amount necessary for combustion, the quantity required where 100 per cent is supplied for dilution, as is the commonly accepted condition in the case of natural draft and hand-firing, will be 24 pounds. But with forced draft the quantity of air required for dilution, as stated by Rankine,[*] " is certainly much less than that which is required in furnaces with chimney-draft; and there is reason to believe that on an

---

[*] " A Manual of the Steam-engine." W. J. M. Rankine.

average it may be estimated at about *one half* of the air required for combustion." That is, the total amount would be 18 pounds.

This applies where hand-firing is the practice. But when, through the action of a properly applied mechanical stoker supplied with air under pressure, as by means of a fan, the bed of fuel is constantly maintained in the most suitable condition for utilizing the air supplied, the amount required for dilution is reduced to a minimum. This is particularly true when the stoker-grate provides special advantages for the equable distribution of the air. Recent tests by Whitham* show not only the decreased air-supply necessary with a good mechanical stoker, but also the reduction in the amount of air required per pound of fuel when a high rate of combustion is maintained by the use of forced draft. With a combustion of twelve pounds of buckwheat coal per square foot of grate per hour, the air was found to be 85.6 per cent in excess of that chemically required; while with a rate of 45.4 pounds almost perfect evaporative efficiency was secured when there was an actual deficiency of 11.2 per cent in the air-supply below the chemical requirements. Startling as this result appears, it is reported by an able expert engineer. It certainly points toward the possibilities of reduced air-supply with mechanical draft.

"In almost all large boiler-furnaces," as stated by Gale,† "a material improvement in economy may be made by cutting down the grate-surface and employing forced draft. Theoretically, 12 pounds of air are sufficient to completely burn a pound of average coal; but in practice, with large grate-surfaces and weak draft, between 20 and 30 pounds are required. By the employment of forced draft and judicious proportioning of the furnace the quantity of air may be easily reduced

---

* "Experiments with Automatic Mechanical Stokers." J. M. Whitham, Trans. Am. Soc. of Mech. Engrs., vol. XVII.

† "Coal as a Source of Power." H. B. Gale. A paper read before the California Electrical Society, May 15, 1893.

to 18 pounds, with the result of a white heat in the furnace and better combustion, besides the saving of a great part of the expense of a high chimney."

As the weight of dry air at 62° is 0.0761 pounds per cubic foot, the volumes corresponding to the above weights are as indicated in Table No. 14.

TABLE No. 14.

AMOUNT OF AIR REQUIRED FOR COMBUSTION.

	Without Dilution.	With 50 per cent Dilution.	With 100 per cent Dilution.
Weight of air................	12 pounds	18 pounds	24 pounds
Volume of air, exact.........	157.7 cu. ft.	236.5 cu. ft.	315.4 cu. ft.
Volume of air, in round numbers	150 cu. ft.	225 cu. ft.	300 cu. ft.

The latter figures, given in round numbers, are those usually employed.

An insufficient supply of air causes imperfect combustion of the fuel, which in bituminous coal is indicated by the production of smoke, and in coke and anthracite coal by the discharge of carbonic oxide from the chimney. An excess of air causes waste of heat to the amount corresponding to the weight of air in excess of that which is necessary, and to the elevation of temperature at which it is discharged from the chimney above that of the external air. Obviously, the maximum efficiency to be secured in the process of combustion is to be sought between these two extremes.

Analysis of Flue-gases.—It is a comparatively simple matter, by means of the proper apparatus, to determine from samples the relative proportions of carbonic oxide, carbonic acid, and oxygen in the gases leaving a boiler-furnace.

The Orsat apparatus,[*] most generally employed for this purpose, is shown in Fig. 2 in its portable form. As will be noted, it consists of three pipettes, P', P'', and P''', in con-

---

[*] Simple instructions for its operation will be found in "Gas and Fuel Analysis for Engineers," by Augustus H. Gill.

nection with a graduated burette, B, for measuring the volumes of gas and a levelling-bottle, A, to control the movement of the gases undergoing analysis. The pipettes contain respectively potassium hydrate for the absorption of carbonic acid, an alkaline solution of potassium pyrogallate to absorb the oxygen, and cuprous chloride to absorb the carbonic oxide. By means of the levelling-bottle the sample of flue-gas is forced through the pipettes in the order named, and the amount absorbed by each is measured by means of the burette. The

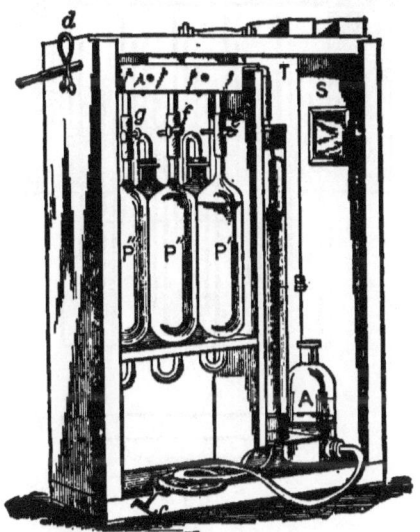

FIG. 2.—ORSAT'S GAS APPARATUS.

amounts by volume thus obtained may be readily transformed into amounts by weight by multiplying by the densities of the various gases. Although the residue is almost entirely nitrogen, it frequently contains small quantities of hydrogen and hydrocarbons, which are more difficult of determination.

This apparatus gives results which are accurate to 0.2 o' one per cent. An analysis with a single apparatus requires about twenty minutes. By making use of two apparatus, and operating them together, two analyses can be made in about twenty-five minutes.

Much depends upon the care with which the gas-samples are collected. When the conditions permit, the arrangement devised by Hoadley and illustrated in Fig. 3 is serviceable. The gases pass from the flue A through $\frac{1}{4}$-inch gas-pipes to the mixing-box B. From this box, indicated as B' in the elevation, they are drawn through four tubes CC to the mixing-chamber D, and thence through E to the aspirator.

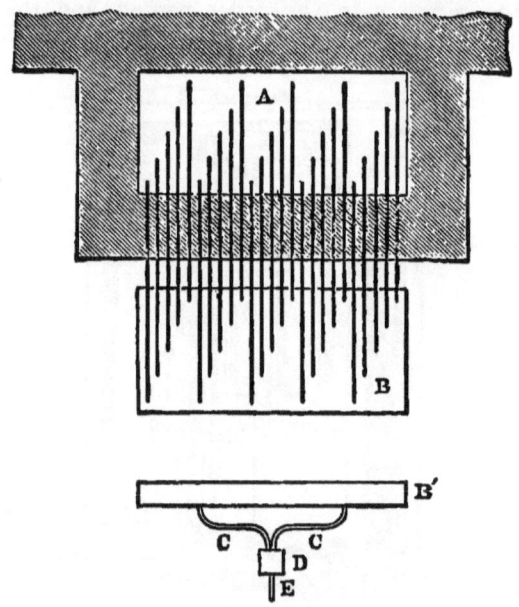

Fig. 3.—Hoadley's Gas-collecting Device.

The aspirator or apparatus for producing suction may take either the form of a water-pump or a steam-pump, relatively very small in its proportions. The simplest apparatus for general use is the jet-pump, depending for its action upon a considerable head of water. A common type is the Richard's jet-pump or aspirator, which in its general construction and operation much resembles an ordinary steam-boiler injector, and is very simply operated by the supply from an ordinary faucet.

By means of the aspirator, acting as in the case of the Hoadley device to produce a partial vacuum in the mixing-chamber D, a uniform flow of gas is induced through each of the tubes. The disposition of the ends of these tubes within the flue or chimney is such as to insure samples being taken from every part. These become thoroughly mixed before passing to the receivers, and being continuously collected are thus rendered average samples of the entire volume passing from the boiler. The mere taking of a sample from a single point, as when a single sample tube is employed, frequently brings about deceptive results. This is likewise true of samples collected only at considerable intervals. Ignorance or disregard of these facts will lead to much wasted effort on the part of the experimenter.

Preparatory to analysis, the gas may be finally collected in a series of sample bottles which, being originally filled with water and hence devoid of air, are gradually emptied and the space filled by the inflowing gas. When everything is ready for analysis water is again caused to flow into the bottle, thereby expelling the gas and forcing it through the gas apparatus. A fair idea of the variety in the composition of the gases from different coals is presented by Table No. 15, in which are given the results of analyses by Bunte. The coals from which the gases were collected were burned in a special experimental apparatus.

Calculation of Air-supply from Gas-analysis.—Disregarding the small proportion of hydrogen and hydrocarbons which may be present in the gases, the residue may be considered as free nitrogen. Knowing the proportion in which it is associated with oxygen to form air, the original amount of air supplied may be calculated from the composition of the gases.

To illustrate the method of calculation, take for instance the result of an analysis showing 11.5 per cent of carbonic acid, 0.9 per cent of carbonic oxide, and 7.4 per cent of free

TABLE No. 15.

COMPOSITION OF GASES FROM DIFFERENT COALS.

	Min. and Max. of Air.	CO_2	CO	H	O	N
Coal from the Ruhr............	....	10.26	0.53	0.01	10.00	79.20
Do.	....	16.45	1.94	1.45	1.52	78.64
Do.	....	13.40	0.48	0.30	6.52	79.30
Do.	....	11.45	1.22	0.78	7.27	79.28
Do. (grate more open).	....	8.15	0.10	0.01	11.60	80.14
Do. Do.	....	6.12	0.89	0.10	14.21	78.68
Coal from Saarbruck: Kœnig..	Min.	15.12	1.09	1.02	2.64	80.13
	Max.	7.07	0.18	0.00	12.57	80.25
" " Trémosna: Bohemia	Min.	13.78	4.69	0.16	1.10	80.27
	Max.	7.94	0.03	0.09	11.03	80.91
" " Hausham: Bavaria.	Min.	10.48	0.07	0.19	9.28	79.98
	Max.	5.71	0.14	0.08	14.86	79.21
" " Miesbach: Bavaria.	Min.	11.46	0.07	0.07	8.66	79.74
	Max.	5 42	0.03	0.02	15.00	79.53
" " Bohemia............	Min.	17.48	1.21	0.06	3.13	78.12
	Max.	12.20	?	0.30	7.87	?
" " the Ruhr: General Erbstolln........	Min.	16.45	1.94	1.45	1.52	78.64
	Max.	3 95	0.06	0.00	16.41	79.58
" " the Ruhr: Gelsenkirchen..........	Min.	10.46	0.11	0.11	8.58	80.74
	Max.	5.44	0.12	0.10	14.15	80.19
" " Saarbruck: Saint-Ingbert.......	Min.	10.73	0.15	0.30	7.36	81.46
	Max.	7.48	0.07	0.10	11.91	80.44
" " Saarbruck: Mittelbexbach........	Min.	13.30	0.61	0.33	4.13	81.63
	Max.	8.44	0.19	0.16	10.58	80.63
" " Saarbruck: Heinitz	Min.	14.62	2 07	1.00	2.07	80.24
	Max.	6.49	0.07	0.06	12.70	80.68
" " Saarbruck: mixed ..	Min.	10.22	0.22	0.07	8.57	80.92
	Max.	8.21	0.04	0.02	10.64	81.09
" " Bohemia............	Min.	15.50	0.74	0.33	1.67	81.66
	Max.	8.48	0.08	0.07	9.69	81.68
" " "	Min.	9.61	0.16	0.08	9.47	80.68
	Max.	7.00	0.11	0.05	12.70	80.14
" " Saxony............	Min.	13.80	0.33	0.30	4.36	81.21
	Max.	7.60	0.16	0.09	11.53	80.62
" " Silesia.............	Min.	11.4	0 15	0.04	7.45	81.22
	Max.	8.07	0.10	0.09	10.73	81.01
" " Bavaria: Peissenberg............	Min.	13.96	1.46	0.79	2.93	80.86
	Max.	7.85	0.07	0.13	10.57	81.38
Lignite from Bohemia....... ..	Min.	14.91	1.04	0.60	2.92	80.53
	Max.	6.36	0.16	0 23	13.15	80.10
Coke from Saarbruck....	Min.	14.87	0.13	0.09	4.16	80.75
	Max.	8.01	0.03	0.00	10.87	81.09

oxygen, all by volume. Evidently, the nitrogen being the only other constituent of the flue-gases which is of importance, it must be present in sufficient quantity to make up the unit volume of gas. Its volume will therefore be

$$100 - (11.5 + 0.9 + 7.4) = 80.2 \text{ per cent.}$$

In the calculation of the weight of nitrogen and of air-supply, it is convenient to treat the percentages by volume as the number of cubic feet of the several gases in 100 cubic feet of flue-gas. Referring to Table No. 11 for the proper volumes, as therein given, the composition of the flue-gas by weight appears to be—

Gas.	Volume.	Density.	Weight.
Carbonic acid	11.5	0.12341	1.4192
Carbonic oxide	0.9	0.07806	0.0703
Oxygen	7.4	0.08928	0.6607
Nitrogen	80.2	0.07837	6.2853

As the atomic weights of carbon and oxygen are respectively 12 and 16, it is evident, as is shown by the following simple calculation, that in one pound of carbonic acid the oxygen constitutes

$$\frac{2 \times 16}{12 + (2 \times 16)} = \frac{32}{44} = \frac{8}{11}$$

of the weight, the remaining $\frac{3}{11}$ being carbon. In a similar manner, it appears that one pound of carbonic oxide is composed of

$$\frac{16}{12 + 16} = \frac{16}{28} = \frac{4}{7}$$

of a pound of oxygen and $\frac{3}{7}$ of a pound of carbon. Therefore the weight of oxygen in 100 cubic feet of the above-stated flue-gases would be—

In the carbonic acid, $\frac{8}{11} \times 1.4192 = 1.0322$ pounds
In the carbonic oxide, $\frac{4}{7} \times 0.0703 = 0.0402$ "
Free oxygen................. 0.6607 "

Total weight of oxygen...... 1.7331 pounds

and the weight of carbon would be—

In the carbonic acid, $\frac{3}{11} \times 1.4192 = 0.3870$ pounds
In the carbonic oxide, $\frac{3}{7} \times 0.0703 = 0.0301$ "
                                                        "

Total weight of carbon ...... 0.4171 pounds

As the air consists, by weight, of 0.2317 parts of oxygen, the above-estimated weight of oxygen would be contained in

$$\frac{1.7331}{0.2317} = 7.48 \text{ pounds of air,}$$

and the supply of air per pound of carbon, the combustion of which resulted in flue-gases having the composition given upon the preceding page, must therefore have been

$$\frac{7.48}{0.4171} = 17.93 \text{ pounds.}$$

If the coal from the combustion of which these gases resulted had contained 85 per cent of carbon, 3.7 per cent of hydrogen, and 2.4 per cent of oxygen, the air-supply per pound of coal would be calculated as follows: The supply of air per pound of coal, disregarding the oxygen and hydrogen present therein, would be

$$0.85 \times 17.93 = 15.24 \text{ pounds.}$$

But on the basis already established, that the oxygen in the fuel renders inert one eighth of its weight of hydrogen, and the remnant is available for combustion, there would be added to the air per pound of coal

$$36\left(0.037 - \frac{0.024}{8}\right) = 0.468 \text{ pounds,}$$

making the total air-supply per pound of coal

$$15.24 + 0.468 = 15.708 \text{ pounds.}$$

Measurement of Air-supply by Anemometer. — The direct method of determining the volume and equivalent weight of air entering the ash-pits by natural or induced mechanical draft consists in ascertaining the velocity by means of an anemometer and multiplying this velocity by the area of the opening. An ordinary form of this instrument is that shown in Fig. 4. It consists of a light and delicately

FIG. 4.—ANEMOMETER.

constructed fan-wheel whose motion is transmitted to a practically frictionless system of gearing within the attached case. The movement of this mechanism is rendered evident by the hands and graduated circles upon the dial, the velocity of the air in feet per minute being indicated thereon by the difference in readings at stated intervals. Such an instrument always requires correction, and should be frequently calibrated to secure reliable results.

As applied for determining the velocity of air entering the ash-pit of a boiler, it should be placed well within a pipe several feet in length, fitted to the ash-door opening, and of an area at least equal to that of the opening. Readings may be taken through glazed openings in the sides of the pipe opposite the anemometer, which should be supported at a predetermined point within the pipe.

It is unlikely with the ordinary boiler that such determination of air-supply will check very closely with that revealed by the gas-analysis because of incidental leakage which may occur through settings and doors, between the point of measurement by the anemometer and the place at which the gas-sample is collected.

Heat of Combustion.—To determine the amount of heat which a given combustible generates requires a somewhat complicated apparatus and considerable care and experience in its operation. Such an apparatus consists of two essential elements, a combustion-chamber and a calorimetric bath within which it is immersed.

Many forms of fuel-calorimeters have been devised, but among the best is that of Mahler.* The combustion-chamber, familiarly known as the "bomb," is of the form illustrated in Fig. 5. The shell is of steel and capable of withstanding great internal pressure. A small known quantity of the combustible to be experimented upon is placed in the tray. The cap S is made tight by means of the lead washer P, and by manipulation of the valve-screw R, oxygen is admitted through N. The bomb is then placed in the bath shown in Fig. 6, containing a known quantity of water, the combustible is ignited by means of an electric current passing through one of the suspension-rods of the tray, and the increase in the temperature of the water in the bath is noted. After due correction there is thus determined the number of thermal units generated during the process of combustion.† The number of British thermal units developed by the complete combustion of a given substance is denominated its heat of

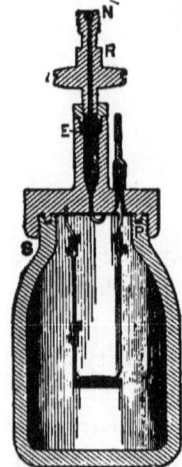

Fig. 5.

---

* The description of this apparatus as given by him before the Société d'Encouragement de Paris in 1892 has been translated and is to be found in " The Calorific Power of Fuels," by Herman Poole.

† For detailed instructions for operation, see " Gas and Fuel Analysis for Engineers," by Augustus H. Gill.

combustion. Evidently the calorific powers of various combustibles as determined by different experimental methods cannot be expected to perfectly agree. In round numbers

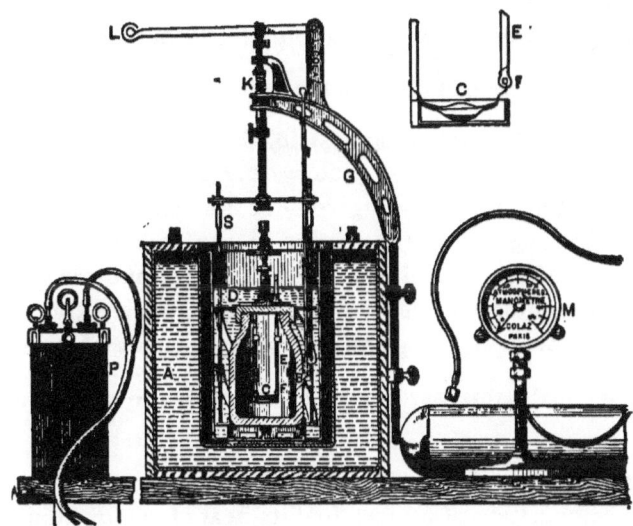

FIG. 6.

the commonly accepted values for the specified substances are as given below.

As determined by the most recent and refined calorimetric tests, the heat of combustion, as measured by the number of British thermal units that are given out upon the combustion of one pound of a given substance, is for each of the following—

Carbon burned to CO_2............ 14,650 B. T. U.
Carbon burned to CO............ 4,400 "
Hydrogen .... 62,100 "
Marsh-gas .... 23,513 "
Olefiant gas.................. 21,343 "
Carbonic oxide burned to CO_2.... 4,393 "

The great loss of heat, due to the incomplete combustion of carbon, is clearly presented in the differences between the total heat of perfect combustion of carbon to CO_2 (viz., 14,650 B. T. U.) and that of carbon to CO (viz., 4400 B. T. U.); the latter being the product of incomplete combustion, as already stated in a previous section.

One pound of carbon, when imperfectly burned, produces $\dfrac{12 + 16}{12} = 2\tfrac{1}{3}$ pounds of carbonic oxide. If this quantity of gas be burned to form carbonic acid, the total amount of heat given out will be $14{,}650 - 4400 = 10{,}250$ B. T. U.; showing that ultimately the carbon gives out its full heat-value no matter what the order of formation of the carbonic acid may have been, whether by direct union of carbon and oxygen, or through the intermediate agency of the carbonic oxide. As the 10,250 B. T. U. are given out by $2\tfrac{1}{3}$ pounds of carbonic oxide, its heat-value per pound is, evidently, $\dfrac{10{,}250}{2\tfrac{1}{3}} = 4393$ B. T. U.

Calculation of the Heat of Combustion.—From the known composition of a given fuel its heat of combustion may in a simple manner be approximately calculated. For such approximation it is customary to disregard that portion of the hydrogen for which there exists in the fuel a sufficient amount of oxygen to form water. The remainder of the carbon and hydrogen thus becomes available for producing heat, and may be introduced in an approximate formula, based upon that for estimating the air required for combustion. Disregarding the effect of inherent nitrogen and sulphur, this formula may be thus expressed:

$$\text{Heat, in B. T. U.,} = 14{,}650\, C + 62{,}100\left(H - \frac{O}{8}\right),$$

in which the weights of carbon, hydrogen, and oxygen in one pound of fuel are respectively represented by their symbols, C, H, and O.

This formula, applied to the determination of the heat of

combustion of the Maryland semi-bituminous coal in Table No. 10, appears as follows:

$$\text{Heat, in B. T. U.,} = 14,650 \times 0.80 + 62,100 \left(0.05 - \frac{0.027}{8}\right)$$
$$= 14,615.$$

Theoretically, the total calorific value, as determined by the calorimeter and as calculated from analysis, should agree. But, on the one hand, there is opportunity for error or imperfection on the part of the calorimeter; while, on the other, the formula employed for calculation from the analysis may fail to make due allowance for heat lost in dissociation, or may not properly recognize the influence of minor constituents. In both cases there is great difficulty in obtaining similar samples. This accounts for differences which frequently exist in reported results. Thus the calorimetric tests of Scheurer-Kestner were about ten per cent, on an average, higher than the analyses; while results reported by Dean* show the calorific value, as determined by calorimeter, to be about six per cent less than that calculated from the test. A comparison of the results obtained by these two methods of determination is presented in Table No. 16, from the tests of Mahler on various American and foreign coals.

In boiler-practice, owing to the opportunities for loss of heat through radiation, heat carried off by flue-gases, incomplete combustion, etc., the maximum efficiency attainable with the best possible boiler and warm-blast or feed-water heating apparatus appears to be about 90 per cent. Under ordinary conditions with good coal the efficiency may be assumed to average from 60 to 70 per cent, and with poor coal from 50 to 60 per cent. It is therefore customary, for rough figuring, to consider the available heat of combustion per pound of fuel to be ordinarily from 10,000 to 12,000 B. T. U.

---

* Transactions Am. Soc. Mech. Engineers, vol. XVII. p. 285.

TABLE No. 16.

HEAT OF COMBUSTION OF FUELS.

Kind of Fuel.	Analysis.					Per cent of Volatile Matter Exclusive of Water and Ash.	Calorific Value Observed. B. T. U.	Calorific Value Excluding Water and Ash. B. T. U.	Calorific Value Calculated. B. T. U.
	Carbon.	Hydrogen.	Oxygen and Nitrogen.	Hydroscopic Water.	Ash.				
Anthracite, from Pennsylvania..	86.456	1.995	2.199	3.450	5.900	3.00	13471	14861	15210
Semi-anth., from Commentry....	84.928	2.892	5.005	1.775	5.400	3.19	14130	15221	15048
Semi-bituminous, from Aniche ..	85.937	4.198	5.240	0.625	4.000	11.93	15167	15901	15638
Bituminous, from Anzin..........	83.754	4.385	5.761	1.100	5.000	21.51	14492	15433	15640
Wigan cannel coal...............	78.382	5.060	5.058	0.600	10.900	31.64	13970	15682	16220
Lignite, from Styria.............	65.455	4.782	24.303	0.710	4.750	50.34	13111	11963	11898
Coke, Pennsylvania anthracite ..	91.036	0.685	2.146	0.233	5.900	68.93	13550	14465	14540

The total heat of various fuels will be shown in succeeding tables.

Ideal Temperature of Combustion.—From the known total and specific heats of combustibles may be calculated the temperature which would result from their combustion if all possible losses were prevented. In ordinary practice those losses must occur and the efficiency of fuels be reduced thereby. It is therefore impossible to attain in practice the full ideal temperature. The general properties of the substances entering into a discussion of the combustion of fuels are given in Table No. 17.

It has already been shown that one pound of carbon, burned to carbonic acid, requires $2\frac{2}{3}$ pounds of oxygen. Hence the total product of combustion of one pound of carbon $= 3\frac{2}{3}$ pounds, as is also evident by the following calculation based upon the atomic weights:

$$\frac{12 + (2 \times 16)}{12} = 3\frac{2}{3} \text{ pounds.}$$

COMBUSTION. 53

TABLE No. 17.

PROPERTIES OF SUBSTANCES CONCERNED IN COMBUSTION.

Substance.	Symbol.	Atomic or Molecular Weight.	Specific Volume.	Specific Heat in a Gaseous Condition.	Density of Weight per Cubic Foot. Pounds.
Hydrogen........	H	1	178.881	3.409	0.00559
Carbon..........	C	12	....	....	....
Nitrogen.........	N	14	12.7561	0.2438	0.07837
Oxygen	O	16	11.2070	0.2175	0.08928
Carbonic oxide....	CO	12 + 16	12.81	0.2450	0.07806
Carbonic acid.....	CO$_2$	12 + 2 × 16	8.10324	0.2169	0.12341
Water...........	H$_2$O	2 + 16	....	0.4805	....
Air.............	......	....	12.3909	0.2375	0.08071
Ash.............	......	....	....	0.2	....

It has further been shown that a total of 11.51 pounds of air is required to furnish 2⅔ pounds of oxygen. Therefore the total weight of the products or results of combustion must be 12.51 pounds, and the weight of the nitrogen alone 12.51 − 3⅔ = 8.84 pounds. As the specific heat of a substance is a measure of the number of thermal units necessary to raise its temperature through one degree, the total number of units required to raise through one degree the products of combustion of one pound of carbon, with the associated nitrogen, may be determined thus:

       Weight. Specific Heat. B. T. U.
Carbonic acid............ 3⅔ × 0.2169 = 0.7953
Nitrogen................ 8.84 × 0.2438 = 2.1552

   B. T. U. per degree 2.9506

As one pound of carbon in the process of burning gives out 14,650 B. T. U., and as it requires 2.9506 B. T. U. to raise through one degree the products of combustion, including the accompanying nitrogen, the ideal temperature resulting from the combustion of one pound of carbon must be 14,650 ÷ 2.9506 = 4965°. In the same manner the ideal temperature of combustion of hydrogen may be calculated, and as it makes no difference in the temperature whether the oxygen required for this union is derived from the original constituents of the fuel or from the atmosphere, the entire amount of hydrogen in the fuel is taken into account.

For the purpose of illustrating the process of approximate calculation, the Maryland semi-bituminous coal in Table No. 10 may be again considered. The important constituents of this coal, expressed in per cent of one pound of coal, are:

    Carbon............................ 80.0 per cent
    Hydrogen .... 5.0 "
    Oxygen............................ 2.7 "
    Nitrogen........................... 1.1 "

For simplicity the ash and sulphur may be disregarded, and also the latent heat of the steam formed by the combination of hydrogen and oxygen. The heat of combustion of this coal has already been calculated as 14,615 B. T. U. By the process explained in the section on " Air Required for Combustion," it has also been shown that, for the total combustion of the carbon and hydrogen contained in one pound of this coal, there are required 10.82 pounds of air, of which 2.51 pounds will be oxygen and 8.31 pounds will be nitrogen. This amount of nitrogen, added to that already in the coal, makes the total $8.31 + 1.1 = 9.41$ pounds.

The total amount of carbonic acid produced by the union of oxygen with 0.8 pounds of carbon is $3\frac{2}{3} \times 0.8 = 2.933$ pounds, and as the total products of combustion of one pound of hydrogen are $\dfrac{2 + 16}{2} = 9$ pounds, the weight of the products of combustion of the hydrogen in the coal will be $9 \times 0.05 = 0.45$ pounds. Hence the thermal units required to raise each of these combustibles through one degree are:

	Weight.	Specific Heat.	B. T. U.
Carbonic acid............	2.933	× 0.2169 =	0.6362
Water...................	0.45	× 0.4805 =	0.2162
Nitrogen................	9.41	× 0.2438 =	2.2942
Total B. T. U............			3.1466

The ideal temperature of combustion, therefore, appears to be

$$14,615 \div 3.1466 = 4645.$$

If, for the purposes of dilution, there had been provided 50 per cent of air in excess of that theoretically required for complete combustion, the amount of heat necessary to raise the temperature of the products of combustion through one degree would have been increased, and the final temperature reduced, as is evident from the following:

	Weight.	Specific Heat.	B. T. U.
50 per cent air for dilution,	$\dfrac{10.82}{2}$ =	5.41 × 0.2375 =	1.2849
Products without dilution as above.		=	3.1466
Total B. T. U.			4.4315

and $14,615 \div 4.4315 = 3296°$. The cooling effect of the air, which is absolutely necessary for dilution, is thus made evident by a decrease of $4645 - 3296 = 1349°$ when it is only 50 per cent in excess.

While the temperature of combustion of a complex fuel may be calculated with much greater refinement by taking into account all of the minor constituents, the results thus obtained are practically of but little more value than those derived from this approximate method; for local conditions in boiler-practice always have considerable effect in reducing the actual temperature to somewhat below the ideal. Hoadley,[*] in carefully conducted tests with a water-platinum calorimeter, found in the heart of the fire under an ordinary boiler a temperature of 2426°, the coal consisting of 82 per cent of carbon, and the supply of air being 21.4 pounds per pound of coal. Immediately above the fire, and at the bridge-wall, the temperature rapidly decreased through losses by radiation and conduction to the walls and the water in the

---

[*] "Warm-blast Steam-boiler Furnace." J. C. Hoadley.

boiler, so that the corresponding temperature at the bridge-wall was only 1341°.

The ideal temperature of combustion of the Maryland semi-bituminous coal, with different degrees of dilution, as determined by calculation in the manner already indicated, is presented in Table No. 18.

TABLE No. 18.

IDEAL TEMPERATURES OF COMBUSTION WITH DIFFERENT DEGREES OF DILUTION.

Percentage of Dilution.	Ideal Temperature.	Loss of Temperature due to Dilution.
0	4645°	....
50	3296	1349
100	2557	2088
150	2087	2558

These figures indicate only the increments of temperature under the given conditions; hence, to obtain the actual thermometric temperature, they must be increased by the initial temperature of the air. Thus, if the air is supplied at 62°, the ideal temperature, with 100 per cent dilution, would be $2557 + 62 = 2619°$, while if the air had been previously heated by special means to 300°, it would be $2557 + 300 = 2857°$.

CHAPTER IV.

FUELS.

Definition.—Fuels may be defined as those substances which, by means of atmospheric air, can be economically burned to generate heat. The principal constituent of all is carbon, with which hydrogen is usually associated. They may be broadly classified as natural and artificial.

Natural Fuels.—Natural fuels are such forms of carbon and its compounds with hydrogen as occur distributed in nature, either as products of existing organic life or as the fossilized remains of a prehistoric growth. Under this heading are included the varieties of wood, coal, mineral oil, and natural gas. The solid fuels may be classified as follows:

WOOD.
PEAT.
COAL.
- Lignite.
- Bituminous.
  - Non-caking, rich in oxygen.
  - Caking.
  - Non-caking, rich in carbon.
- Anthracite.

Artificial Fuels.—Artificial fuels comprise those forms of carbon or its compounds with hydrogen which owe their origin to some process of manufacture, but are not commonly found distributed in nature. These are generally obtained from natural fuels by some special process; as, for instance, charcoal from wood, coke and volatile hydrocarbons from coal. Artificial fuels include the various attempts to cement together, in the form of blocks or briquettes, such combustible refuse as is too small to be otherwise profitably consumed.

The products of carbonization may be classified as follows:

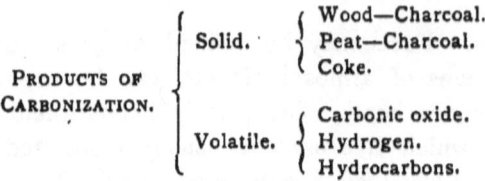

Wood.—Although the term "wood" broadly includes all substances of vegetable fibre which have not undergone geological changes, it applies directly to the fairly compact substance which constitutes tree trunks and branches. With reference to its heating-power, wood under this definition may be classed as hard and soft. Hard woods include the oak, hickory, maple, beech, and walnut; and soft woods, the pine, elm, birch, chestnut, poplar, and willow. When freshly cut, wood contains nearly 50 per cent of moisture, which seriously reduces its calorific value. Through the process of air- or kiln-drying the amount of moisture may be brought down, to from 10 to 20 per cent.

The average chemical composition of different kinds of wood and their calorific values are given in Table No. 19.

Straw and Tan.—Evidently straw can be economically employed as a fuel only where the supply is directly at hand and the cost of other fuel is excessive.

Straw of average composition weighs, when pressed, 6 to 8 pounds per cubic foot. Oak bark, after having served its purpose as a tanning agent, thereby becoming spent tan and consisting only of the fibrous portion of the bark, is used as a fuel, but only under the economical conditions which hold in the use of straw as a fuel; that is, when the tan is readily accessible and its total cost when placed in the furnace is less, for a given result, than that of other available fuels. In the process of tanning the bark loses about 20 per cent of its weight. Perfectly dry tan, containing 15 per cent of ash, has a heating-power of 6100 B. T. U.; while tan containing 30

Table No. 19.

COMPOSITION AND CALORIFIC VALUE OF WOOD, TAN, AND STRAW.

Name.	Carbon.	Hydrogen	Oxygen.	Nitrogen.	Ash.	Water.
Beech................	49.36	6.01	42.69	0.91	1.06	
Oak..................	49.64	5.92	41.16	1.29	1.97	
Birch................	50.20	6.20	41.62	1.15	0.81	
Poplar...............	49.37	6.21	41.60	0.96	1.86	
Willow...............	49.96	5.96	39.56	0.96	3.37	
Average for wood.....	49.70	6.06	41.30	1.05	1.80	
Wheat straw..........	35.86	5.01	37.68	0.45	5.00	16.00
Barley "	36.27	5.07	38.26	0.40	4.50	15.50
Mean for straws......	36.06	5.04	37.97	0.42	4.75	15.75

Calculated by formula already presented, the calorific value of wood of the average composition above given is 7838 B.T.U.

Perfectly dry tan containing 15 per cent of ash has a calorific value of 6100 B.T.U., while that of tan containing 30 per cent of water is 4284 B.T.U.

Straw of mean composition given above has a calorific value, deducting the heat lost in evaporating its constituent water, of 5155 B.T.U.

per cent of water—its usual condition of dryness—has a calorific value of only 4284 B. T. U.

The composition and calorific power of straw and tan are given in Table No. 19.

The conditions of success in burning tan, as is the case with all wet fuel, consist in completely surrounding it with heated surfaces and burning fuel so that it may be rapidly dried, and then so arranging the apparatus that thorough combustion may be secured.

Bagasse.—The term "bagasse," or megass, is generally understood to apply to that portion of the sugar-cane which is left after extracting the juice. As the methods of extraction give results varying all the way from 40 per cent to 80 per cent, it is evident that it includes substances differing greatly in composition. In its broadest sense it may, therefore, be taken as meaning the refuse discharged from the cane-mill or diffusion process, whether it comes from a mill giving 40 per cent extraction and leaving 70 per cent of moisture, or whether it be the air-dried bagasse of the tropics with only 10 per cent of moisture.

Mill-bagasse is the refuse left after the juice has been extracted by means of the mill-rolls. Diffusion-bagasse is the material remaining after a series of soaking processes for which it has been chopped into small pieces, and whereby the saccharine matter has diffused itself throughout the mass of water in which the cane has been placed. The original cane, and likewise the bagasse, consist of woody fibre, water, and combustible salts; but, naturally, the squeezing process reduces the percentage of liquid matter, and proportionately increases the relative amount of fibrous material in the bagasse. Upon the fibre, which is principally carbon, the value of bagasse as a fuel largely depends. In tropical canes it constitutes in round numbers about 12 per cent of the original cane, while in Louisiana cane 10 per cent is a fair average. Tropical cane and the bagasse therefrom have the composition given in Table No. 20.

FUELS. 61

TABLE No. 20.
COMPOSITION OF TROPICAL CANE AND BAGASSE.

Constituents.	Cane.	Bagasse.		
		66 per cent Extraction.	70 per cent Extraction.	72 per cent Extraction.
Woody fibre	12.5	37	40	45
Water	73.4	53	50	46
Combustible salts	14.1	10	10	9

The proportional composition of Louisiana bagasse is clearly shown, for different degrees of extraction, in Table No. 21.

TABLE No. 21.
COMPOSITION OF DRY LOUISIANA BAGASSE.

Constituents.	Percentage.
Volatile matter	81.37
Fixed carbon	14.26
Ash	4.6

Carefully conducted calorimetric tests of bagasse, when under different conditions, by Atwater, give the heat-values which are indicated in Table No. 22.

TABLE No. 22.
CALORIMETRIC TESTS OF BAGASSE.

Description.	Per cent Moisture in Sample.	B. T. U. per Pound as Received.	B. T. U. per Pound. Dry Matter from Preceding Column.	B. T. U. per Pound. Dry Matter by Actual Test.
Purple cane exhaust-chips, direct from diffusion-battery	90.36	799	8288	8320
Striped cane exhaust-chips, direct from diffusion-battery	90.54	873	9229	8289
Purple cane exhaust-chips, passed through laboratory-mill	73.34	1966	7373	8309
Striped cane exhaust-chips, passed through laboratory-mill	69.62	2547	8384	8384
Averages			8319	8325

This table serves to show the serious effect of contained water upon the heat-value of the bagasse of different extractions. This effect is to be expected, for all of the water in the bagasse must when employed as fuel be vaporized before the combustible matter can be consumed, and in the process of vaporization an enormous amount of heat is rendered latent, and thus lost to the furnace so far as heating effect is concerned.

The only method available in estimating the fuel-value of the different extractions of mill-bagasse is that based upon the assumption that bagasse consists of two substances, fibre and juice, and that this juice has the same composition as that which has already been extracted. To obtain a heat-value for juice, it must be divided into sugar and molasses. Thus, for instance, an average cane consisting of—

Fibre............................	10	per cent
Juice { Sucrose.......................	12	"
Glucose.......................	2	"
Solids—not sugar........	1	"
Water.........................	75	"
	100	per cent,

will, upon passing through a mill giving an extraction of 75 per cent, be reduced to bagasse, the weight of which will be only 25 per cent of the original cane of which it formed a part.

The proportional composition in per cent of the weight of the original cane and of the resulting bagasse will then be as presented in Table No. 23.

Calorimetric tests of molasses, sugar, and fibre indicate the following values:

Molasses........................,6956	B.T.U.
Sugar...........................,7223	"
Fibre............................,8325	"

FUELS. 63

TABLE No. 23.

COMPOSITION OF MILL BAGASSE.

Constituents.	In per cent of Original Cane.	In per cent of the Resulting Bagasse.
Water........................	12.75	51
Fibre........................	10.0	40
Sugar........................	1.5	6
Molasses (dry matter only)	0.75	3
	25.00	100

In Table No. 24 the total heat-values are based upon those of the constituents given above, and a fair sample of Pennsylvania coal having a heat-value of 14,000 B. T. U. is taken as the basis of comparison with coal.

TABLE No. 24.

VALUE OF ONE POUND OF MILL BAGASSE AT DIFFERENT EXTRACTIONS UPON CANE OF 10 PER CENT FIBRE AND JUICE OF 15 PER CENT TOTAL SOLIDS.

Per cent Extraction on Weight of Cane.	Per cent Moisture, in Bagasse.	Fibre.		Sugar.		Molasses.		Total Heat Developed B. T. U.	Heat Required to Evaporate the Water Present. B. T. U.	Heat Available. B. T. U.	Pounds Bagasse required to equal 1 lb. Coal of 14,000 B. T. U. Calorific Power.	Coal Equivalent per Ton of Cane. Pounds.	Temperature of Fire. Fahrenheit.
		Per cent in Bagasse.	Fuel Value. B. T. U.	Per cent in Bagasse.	Fuel Value. B. T. U.	Per cent in Bagasse.	Fuel Value. B. T. U.						
90	0.00	100.00	8325	.....	.....	.....	.....	8325	.....	8325	1.68	119	2465°
85	28.33	66.67	5550	3.33	240	1.67	116	5900	339	5561	2.52	119	2236
80	42.50	50.00	4162	5.00	361	2.50	174	4697	509	4188	3.34	120	2023
75	51.00	40.00	3330	6.00	433	3.00	209	3972	611	3361	4.17	120	1862
70	56.67	33.33	2775	6.67	482	3.33	232	3489	679	2810	4.98	120	1732
65	60.71	28.57	2378	7.15	516	3.57	248	3142	727	2415	5.80	121	1612
60	63.75	25.00	2081	7.50	541	3.75	261	2883	764	2119	6.61	121	1513
55	66.12	22.22	1850	7.78	562	3.88	270	2682	792	1890	7.40	121	1427
50	68.00	20.00	1665	8.00	578	4.00	278	2521	815	1706	8.21	122	1350
45	69.55	18.18	1513	8.18	591	4.09	284	2388	833	1555	9.00	122	1284
40	70.83	16.67	1388	8.33	601	4.17	290	2279	849	1430	9.79	123	1222
25	73.67	13.33	1110	8.67	626	4.33	301	2037	883	1154	12.13	124	1077
15	75.00	11.77	980	8.82	637	4.41	307	1924	899	1025	13.66	124	1002
0	76.50	10.00	832	9.00	650	4.50	313	1795	916	879	15.93	126	906

In somewhat abbreviated form the fuel-values of one pound of diffusion-bagasse, at various degrees of moisture, are given in Table No. 25.

Table No. 25.

FUEL VALUES OF ONE POUND OF DIFFUSION BAGASSE AT VARIOUS DEGREES OF MOISTURE.

Moisture in Bagasse. Per cent.	Heat Developed per Pound of Bagasse. B. T. U.	Heat Available per Pound of Bagasse. B. T. U.	Number of Pounds of Bagasse Equivalent to 1 lb. of Coal.	Estimated Temperature of Fire. Fahrenheit.
0	8325	8325	1.68	2465°
20	6660	6420	2.18	2294
30	5827	5468	2.56	2186
40	4995	4516	3.10	2049
50	4162	3563	3.93	1870
60	3330	2611	5.41	1627
70	2497	1658	8.44	1281
75	2081	1183	11.90	1045

Peat. — Intermediate between wood and coal may be placed peat, which is the result of one of the most important geological changes now in progress. In certain swampy regions in the temperate latitudes there occur immense quantities of semi-aquatic plants, which, under special conditions of heat and moisture, are undergoing a curious chemical transformation, whereby the oxygen of the plant is eliminated, leaving behind as peat a spongy carbonaceous residue. This is found in beds varying from 1 or 2 to 40 feet in depth. That near the surface, which is in a less advanced state of decomposition, is light, spongy, and fibrous, of yellow or light reddish-brown color; lower down it is more compact, and darker in color; while in the lowest strata the color is almost black, and the peat is pitchy and unctuous, with scarcely any evidence of the fibrous texture which exists in the higher strata and in the original vegetable matter from which it was formed.

In its natural condition, peat generally contains from 75 to 80 per cent of its entire weight of water, occasionally amounting to 85 or even 90 per cent. Evidently it is totally unfitted for use as a fuel until it has been dried. By the process of drying it shrinks very decidedly, its specific gravity, when dry, varying from 0.22 or 0.34 for the surface peat, which is light and porous, to 1.06 for the lowest peat, which is very dense. Owing to the abundance of other fuels, peat has been but little used in this country; but in Europe it has already found an extensive field, not only in domestic but in metallurgical and other operations.

In Table No. 26 are given the calorific values and the composition of Irish peats, both exclusive and inclusive of moisture.

Coal.—The extensive distribution, the portable character, and the heat-value of coal make it the principal fuel of all civilized nations. Coal is in effect the reservoir of the stored energy of the sun, by the action of whose heat-rays it was produced. It is a fossil fuel for whose existence geology thus accounts: During that period of the earth's formation known as the carboniferous age, vegetation was rank in the extreme. The atmosphere contained an amount of carbonic acid far in excess of that now present. The presence in the atmosphere of this excess of carbon, which is the food of the plants, as well as the temperature and the climatic conditions, were all favorable to the most prolific development of plant-life. Age after age was employed by this vegetable growth in freeing the atmosphere from carbonic acid, and in storing up the potential energy of the sunlight as woody fibre in the form of carbon, separated from oxygen. By this continuous process of growth and death of vegetable matter the earth became strewed with the remains, which were gradually compacted into peat-beds of enormous extent. With succeeding climatic and geological changes, these peat-beds, one after another, became submerged and overlaid by thousands of feet of sandstone, limestone, and slate. Under the tremendous pressure thus exerted the peat-beds were compressed and converted by successive stages into lignite, brown coal, gaseous coal, bituminous coal, and semi-anthracites.

It must be obvious that sharp lines of demarcation between the various kinds of coal cannot exist, and, therefore, that

TABLE No. 26.
COMPOSITION AND CALORIFIC VALUE OF IRISH PEAT.

EXCLUSIVE OF MOISTURE.

Description.	Moisture.	Carbon.	Hydrogen.	Oxygen.	Nitrogen.	Sulphur.	Ash.	Coke.
Good air-dried.........	......	59.7	6.0	31.9		......	2.4	
Poor air-dried.........	......	59.6	4.3	29.8		......	6.3	
Dense, from Galway....	......	59.5	7.2	24.8	2.3	0.8	5.4	44.3
Averages...............	......	59.6	5.8	29.6		0.3	4.7	

INCLUSIVE OF MOISTURE.

Good air-dried.........	24.2	45.3	4.6	24.1		......	1.8	
Poor air-dried.........	29.4	42.1	3.1	21.0		......	4.4	
Dense, from Galway....	29.3	42.0	5.1	17.5	1.7	0.6	3.8	31.3
Averages...............	27.8	43.1	4.3	21.4		0.2	3.3	

The calculated calorific value of dry Irish peat of the average composition given above is 10,040 B.T.U.

they can only be approximately classified, for one form merges into another. A fair illustration of the different stages in the process of alteration of wood-fibre into anthracite coal is presented in Table No. 27.

TABLE No. 27.

CONVERSION OF WOOD-FIBRE INTO ANTHRACITE.

Description.	Carbon.	Hydrogen.	Oxygen.
Wood-fibre (cellulose).....	52.65	5.25	42.10
Peat.....................	60.44	5.96	33.60
Lignite..................	66.96	5.27	27.76
Lignite (brown coal)......	74.20	5.89	19.90
Coal (bituminous).........	76.18	5.64	18.07
Coal (semi-anthracite)....	90.50	5.05	4.40
Anthracite	92.85	3.96	3.19

Coals are classified according to the amounts of carbon and volatile matter which are present in their composition, although different methods are adopted by different authorities. The following is the classification * generally adopted, beginning with those containing the greatest proportion of carbon:

ANTHRACITES....................... { Hard anthracites.
{ Semi- or gaseous anthracites.

COMMON BITUMINOUS COALS..... { Semi-bituminous coals .... { Semi-bituminous cherry coal.
{ Semi-bituminous splint coal.
{ Bituminous coals. { Caking coal.
{ Cherry coal.
{ Splint coal.

HYDROGENOUS OR GAS COALS .... { Cannel coal.
{ Hydrogenous shaly coal.
{ Asphaltic coal.

LIGNITE.

The general composition of these coals has already been given in Table No. 10. In the consideration of their characteristics they will be taken up in the order of their geological

---

* "Geology of Pennsylvania." H. D. Rogers.

formation. Their progressive alteration from wood to coal is thus clearly indicated.

Lignite.—Although classed among mineral coals, from a geological standpoint lignite properly occupies a position between peat and bituminous coal. It is believed to be of later origin than bituminous coal, and is in a less advanced stage of decomposition. The woody fibre and vegetable texture of lignite are almost entirely wanting in coal, although there is little question as to their common origin. Although much like brown coal in general appearance, lignite differs from it in the fact that upon distillation it yields acetic acid, while brown coal produces only ammoniacal liquor. Like peat, lignite presents much variety in appearance, some specimens being almost as hard as true coal, while others possess a distinctly woody structure and are of a light-brown color. It has an uneven fracture and a dull and somewhat fatty lustre. Being easily broken, it will not readily bear transportation, while exposure to the weather causes it to rapidly absorb moisture and to crumble easily. Its value as a fuel is, therefore, limited, for it must be used near its place of occurrence, and very soon after it is mined. It is non-caking and yields but moderate heat, being inferior to even the poorer varieties of bituminous coal. In this country its use is decidedly limited, being restricted to the locality of the mines which produce it. It is plenteous, however, west of the Mississippi, in which territory it is used to a considerable extent. The three analyses which are presented in Table No. 28 give the

TABLE No. 28.

COMPOSITION OF LIGNITE.

Locality.	Specific Gravity.	Fixed Carbon.	Volatile Combustible Matter.	Water.	Ash.	Total Volatile Matter.	Coke.	Authority.
Kentucky	1.201	40.0	23.0	30.0	7.0	53.0	47.0	Cox.
Washington	....	52.85	31.75	7.00	3.00	38.75	£1.25	"
Colorado	1.271	41.25	46.00	3.50	9.25	49.50	ɛc.50	King.

average composition of samples from the widely separated States of Kentucky, Washington, and Colorado.

Bituminous Coal.—The classification of bituminous coal is rendered difficult because of the lack of definite lines of demarcation between the varieties. As a rule, however, coal containing as much as 18 to 20 per cent of volatile combustible is called bituminous. Some bituminous coal yields, upon analysis, as much as 50 per cent of volatile matter and sometimes more. In *proximate composition*—namely, in fixed carbon, volatile matter, and earthy matter—the bituminous coals may be regarded as ranging between the following general limits:

```
Fixed carbon............. 52 to 84 per cent
Volatile matter.......... 12 to 48    "
Earthy matter............  2 to 20    "
Sulphur..................  1 to  3    "
```

The amount of water expelled by heating to 212° is from 1 to 4 per cent.

In *ultimate composition*, as shown by refined analysis, the approximate range of composition is as follows:

```
Carbon................ 75 to 80 per cent
Hydrogen..............  5 to  6    "
Nitrogen..............  1 to  2    "
Oxygen................  4 to 10    "
Sulphur...............0.4 to  3    "
Ash...................  3 to 10    "
```

In its external properties, ordinary bituminous coal varies in color from a pitch-black to a dark brown, with a lustre that is vitreous or resinous in the more compact specimens, and silky in those showing traces of vegetable fibres. Irrespective of natural joints, the fracture of bituminous coal is generally conchoidal. The distinctive characteristic of this fuel is the emission of yellow flame and smoke when burning.

All bituminous coals may be classified on broad lines as either caking or non-caking.

Caking Coal is the name given to any coal which, when heated, seems to fuse together and swell in size, becomes pasty in appearance, and emits a sticky substance over the surface, while liberating small streams of gas which burn with a bright yellow or reddish flame terminating in smoke. It is characteristic of caking coal that the pasty lumps will cohere in the fire and form spongy-looking masses, not infrequently covering the entire surface of the grate. Such coals, unusually rich in volatile hydrocarbons, are considered most valuable for gas-manufacture.

Non-caking Coal has the property of burning freely in the fire; hence the common appellation, "free-burning coal." The heat does not cause the lumps to fuse or run together. The block coal of the Western States is a representative non-caking coal. It consists of successive layers which are easily separated into thin slices. The surfaces which are thus displayed are generally covered with a layer of very finely divided fibrous carbon and are dull and lustreless. When coal of this character is broken at right angles to this lamination the surface is bright and glistening.

The ultimate composition of various bituminous coals is given in Table No. 10, which has already been presented; while among the coals listed in Table No. 31, which follows on succeeding pages, are also many that would be classed as bituminous.

Cannel Coal is a variety of bituminous coal very rich in carbon. It kindles readily, burns without melting, and emits a bright flame like that of a candle. It differs greatly in appearance from all other bituminous coals, being very homogeneous, having a dull, resinous lustre, and breaking without following any distinct line of fracture. It is exceedingly valuable as a gas-coal because of its richness in hydrocarbons, but is little used in this country as a steam- or boiler-coal.

FUELS. 71

The proximate analysis of a few typical American specimens is presented in Table No. 29.

TABLE No. 29.
COMPOSITION OF CANNEL COAL.

Locality.	Specific Gravity.	Fixed Carbon.	Volatile Matter.	Earthy Matter.	Authority.
Franklin, Pa............	....	40.13	44.85	15.02	Johnson.
Dorton's Branch, Ky....	1.25	55.1	42.9	2.0	Owen.
Breckenridge, Ky.......	....	32.0	55.7	12.3	Peters.
Davis County, Ind......	1.23	42.0	52.0	6.0	Cox.

Semi-bituminous Coal is softer and contains more volatile matter than true anthracite coal, but in its general characteristics closely approaches that fuel. It resembles in appearance the anthracites more closely than it does the bituminous coals, but its fracture is less conchoidal than that of the former; it is lighter and both kindles and burns more rapidly. Because of this latter feature it is extremely valuable as a fuel, for when burned it readily gives off a great quantity of heat and can always be relied upon to keep up an intense and free-burning fire requiring comparatively little attention, is readily cleaned and kept in good condition. It is, when pure, almost entirely free from smoke and soot.

The proximate analysis of semi-bituminous coal from Cumberland, Md., and Blossburg, Pa., is given in Table No. 30.

TABLE No. 30.
COMPOSITION OF SEMI-BITUMINOUS COAL.

Locality.	Specific Gravity.	Fixed Carbon.	Volatile Matter.	Sulphur.	Earthy Matter.	Authority.
Cumberland, Md.....	1.41	68.44	17.28	0.71	13.98	Johnson.
Blossburg, Pa.......	1.32	73.11	15.27	0.85	10.77	"

Semi-anthracite Coal.—Among the semi-anthracite coals are classed those which contain from 7 to 8 per cent of volatile combustible matter. Because of the presence of this ingredient, which apparently exists in the gaseous state in the cells or cracks of the coal, this variety kindles more readily and burns more rapidly than hard anthracite. Analysis of Wilkesbarre, Pa., semi-anthracite,* which is compact, conchoidal, iron-black, and shiny, shows the following to be its composition:

 Fixed carbon .... 88.90 per cent
 Volatile matter.................. 7.68 "
 Earthy matter .... 3.49 "
                                   _____
                                   100.07

Its specific gravity is 1.4.

Anthracite Coal.—Pure anthracite, sometimes called blind coal, ignites slowly, is a poor conductor of heat, and burns at a very high temperature. When pure it consists of—

 Carbon....................... 90 to 94 per cent
 Hydrogen..................... 1 to 3 "
 Oxygen and nitrogen.......... 1 to 3 "
 Water........................ 1 to 2 "
 Ash.......................... 3 to 4 "

It is thus evident that it is composed almost entirely of carbon; in fact, this is its distinguishing characteristic. The hydrocarbons, as evidenced in the volatile constituents, are present in very small proportion. As a consequence, it is not a long-flaming coal, but when in a state of incandescence its radiant power is great, owing to the intensity of combustion of the practically pure carbon of which it consists.

In the process of burning it neither swells, softens, nor gives off smoke. The flame is quite short, of a yellowish tinge, changing to a faint blue, and largely due to the presence of water which is decomposed by the heat. This flame is free from particles of solid carbon, and has the appearance

---

* Geological Survey of Pennsylvania.

of being transparent. Anthracite coal is homogeneous in structure; its fracture is decidedly conchoidal, and it is but slightly affected by exposure to the weather.

Analysis of anthracite coal from Tamaqua, Pa.,* shows it to consist of—

Carbon.......................... 92.07 per cent
Volatile matter................... 5.03   "
Earthy matter.................... 2.90   "
                                 _____
                                 100.00

and to have a specific gravity of 1.57.

Geographical Classification.—Although widely distributed throughout the United States, the various kinds of coal may be geographically classified in a general manner, as follows:

ANTHRACITES .... { Eastern portion of Allegheny Mountains and Rocky Mountains in Colorado.

BITUMINOUS COALS. { Caking.........Mississippi Valley.
Non-caking....Maryland and Virginia.
Cannel.........Pennsylvania, Indiana, and Missouri.

LIGNITES ....Colorado, Kentucky, and Washington.

In Table No 31 † are presented the results of numerous analyses of American coals and their calorific values. Of necessity the ultimate value of any coal as a steam producer must be measured by the amount of water it can evaporate when properly burned in the furnace of a steam-boiler, rather than by the number of thermal units generated by its combustion under experimental conditions.

Petroleum.—The only natural liquid fuel is crude petroleum oil. This is distinctly a hydrocabron liquid, and is found in abundance in certain localities in America and Europe. The principal sources of supply are, however, in the Ohio Valley of the United States, and on the borders of the Caspian Sea in Eastern Europe. It is found in natural cavities

---

* Geological Survey of Pennsylvania.
† Compiled from "Mineral Resources of the United States" and similar publications by coal-producing States.

TABLE No. 31.
COMPOSITION AND CALORIFIC VALUE OF AMERICAN COALS.

Name or Locality.	Constituents in Per Cent of Total Weight.					Fuel Value per Pound of Coal.		
	Moisture.	Volatile Matter.	Fixed Carbon.	Ash.	Sulphur.	B.T.U. Calculated.	B.T.U. by Calorimeter	Theoretical Evaporation in lbs., from and at 212°.

ALABAMA.

Name or Locality.	Moisture	Volatile Matter	Fixed Carbon	Ash	Sulphur	B.T.U. Calc.	B.T.U. Calor.	Theor. Evap.
Brierfield, Bibb Co............	0.935	41.04	55.76	3.20	1.01	14620		15.13
Coalburg, Jefferson Co........	0.83	30.745	65.075	3.014	1.203	14513		15.02
Corona, Walker Co............	1.66	41.12	50.69	7.36		13851		14.34
Patton Junc., Walker Co......	1.17	22.12	68.34	6.00	1.85	13410		13.88
Warrior, " "		33.88	63.03	1.92	1.20	14550		15.06

ARKANSAS.

Coal Hill, Johnson Co.........	1.35	14.93	74.06	9.66	3.04	13713		14.1
" " "	1.70	14.60	74.91	8.79	3.04		11812	12.22
" " "	1.017	10.841	76.119	8.351	3.672	13053		13.51
Felker Slope, Franklin Co.....	1.128	13.211	81.277	3.22	1.164	13927		14.41
Huntington Co................	1.30	18.95	71.51	8.24	0.78		11756	12.17
" "	1.30	18.90	73.15	6.65	0.75		11907	12.32
" "	1.27	18.89	71.74	8.10	0.65		12537	12.97
Huntington Slope, Sebastian Co	0.928	15.546	77.538	4.845	1.143	13813		14.30
Jenny Lind, " "	1.26	17.64	72.48	8.62	2.11	13964		14.4
Lignite.......................						9215		9.54
Philpot Shaft, Johnson Co.....	0.869	14.133	80.915	3.09	0.993	14020		14.51
Quito, Polk Co................	0.980	12.20	76.817	8.174	1.829	13117		13.27
Spadra, Johnson Co............	1.47	13.27	78.63	6.63	1.60	14420		14.9

CALIFORNIA.

Livermore, Alameda Co........	18.08	39.30	35.61	7.01		11438		11.84

FUELS. 75

COLORADO.

Name or Locality.	Constituents in Per Cent of Total Weight.					Fuel Value per Pound of Coal.		
	Moisture.	Volatile Matter.	Fixed Carbon.	Ash.	Sulphur.	B.T.U. Calculated.	B.T.U. by Calorimeter	Theoretical Evaporation in lbs., from and at 212°.
Boulder Co.	12.01	35.19	46.24	6.56		12271		12.70
Fremont Co.	3.93	42.43	47.16	6.48		13555		14.03
" "		42.40	53.72	3.20		14559		15.07
Las Animas Co.	1.26	36.40	53.10	0.924		13434		13.94
Lignite.						13560		14.04
" slack.	14.80	32.00	42.86	10.34	0.76	13865		14.35
" slack, North Colorado	18.88	31.74	40.08	9.30	0.61	8500		8.80
Rouse Mine.	3.13	37.32	30.00	8.25				

ILLINOIS.

Barclay, Sangamon Co.	10.80	27.32	44.78	17.10		10699		11.08
Big Muddy, Jackson Co.	7.39	28.28	53.87	10.46	0.98		11466	11.87
" "	6.12	30.95	53.74	9.19	1.22		11529	11.93
" "	5.85	31.84	55.72	6.59	2.92		11781	12.19
" "	6.35	31.50	55.25	6.90	2.02			13.0
Blair Bluff, Henry Co.	12.60	28.96	48.54	9.90		12567		11.99
Bloomington, McLean Co.	7.90	34.02	53.12	4.96		11578		13.56
Bureau Co.						13100		13.48
Carbondale.	6.36	26.40	59.84	7.40		13025		13.29
Catlin, Vermillion Co.	7.80	31.08	48.42	12.70		12838		12.29
Colchester.	11.60	25.02	44.76	18.62		11871	9848	10.19

ILLINOIS.—Continued.

Name or Locality.	Moisture.	Constituents in Per Cent of Total Weight.				Fuel Value per Pound of Coal.		
		Volatile Matter.	Fixed Carbon.	Ash.	Sulphur.	B.T.U. Calculated.	B.T.U. by Calorimeter	Theoretical Evaporation in lbs., from and at 212°.
Colchester slack.................	5.30	25.45	38.15	31.10	1.20		9035	9.35
Collinsville, Madison Co.........	9.89	45.89	31.57	13.34	5.34		10143	10.50
Dumferline slack.................	9.64	28.86	39.48	22.02			9401	9.73
Duquoin, Perry Co...............	8.86	23.54	60.60	7.00		12697		13.14
Duquoin Jupiter, Perry Co........	11.30	30.31	49.91	8.48	0.91	12175	10710	11.08
Ellsworth, Macoupin Co..........	9.26	42.22	42.17	6.35	2.62	12348		12.60
Elmwood, Peoria Co.............	7.60	27.60	55.30	9.50		11683		12.78
Farmington, Fulton Co............	8.52	29.28	50.48	11.72				12.09
Gillespie, Macoupin Co...........	12.61	30.58	45.27	11.54	1.45		9739	10.09
Girard, " "	9.70	34.39	45.76	10.15	3.49		9954	10.30
" " "	8.90	32.25	42.89	15.96	8.10		10269	10.63
Grape Creek, Vermillion Co......	9.74	28.34	51.32	10.60		11876		12.29
Heitz Bluff, St. Clair Co..........	8.95	37.81	48.24	5.00	3.27	11723		10.69
Johnson's, " " "	5.50	40.14	40.53	13.83	4.80	11552	10332	12.10
Kewanee........................	15.60	27.60	49.66	7.14		11036		11.96
Lincoln, Logan Co...............	10.92	27.60	46.64	14.84		12453		11.42
Lombardville, Stark Co...........	9.42	31.38	51.74	7.46		11479		12.89
Loose's, Sangamon Co............	10.71	37.62	45.07	6.60	2.39	13123		11.9
Mercer Co.......................						12659		13.58
Montauk Co.....................	6.37	31.93	59.13	1.81	0.76	13673		13.10
Mount Carbon, Jackson Co.......	10.38	36.68	46.10	6.84	3.53	17763		14.15
Mount Olive, Macoupin Co.......								12.2
Oakland, St. Clair Co............	8.30	34.40	43.12	14.18	4.42		10395	10.76
Oglesby, La Salle Co.............	12.12	30.84	49.32	7.72		12013		12.44

FUELS.

ILLINOIS.—*Continued.*

Name or Locality.	Constituents in Per Cent of Total Weight.					Fuel Value per Pound of Coal.		
	Moisture.	Volatile Matter.	Fixed Carbon.	Ash.	Sulphur.	B.T.U. Calculated.	B.T.U. by Calorimeter	Theoretical Evaporation in lbs. from and at 212°.
Reinecke, St. Clair Co............	7.56	39.81	42.49	10.14	4.02	11720	10080	12.1
Riverton, Sangamon Co............	11.06	37.94	42.98	8.02	3.27	11406	9261	11.8
St. Bernard......................	14.36	30.86	48.39	6.39	1.38		10080	10.44
St. Clair........................	7.80	30.69	39.68	21.83	9.62		9261	9.58
" "	10.25	33.10	41.79	14.86	6.92		10294	10.65
" "	11.15	34.19	44.94	9.72	4.27		10647	11.02
St. John, Perry Co...............	9.82	28.35	45.77	16.08	2.06		9765	10.10
" " "	13.60	24.46	43.54	15.40	1.83		9828	10.18
Streator, La Salle Co............	12.01	35.32	48.78	3.89	2.38		11403	11.80
Trenton, Clinton Co..............	13.34	30.39	51.96	4.31	0.92		10584	10.96
Vulcan Nut, St. Clair Co.........	9.95	31.04	52.03	6.98	1.04		11245	11.63
" " " "	7.44	30.86	45.09	16.61	1.32		9450	9.78
" " " "	10.30	27.91	48.99	12.80	0.71		10626	11.00

INDIAN TERRITORY.

Atoka............................	6.66	35.42	51.32	6.60	3.73		11088	11.47
Choctaw Nation...................	1.59	23.31	66.85	8.25	1.18		12789	13.23
Lehigh, Atoka Co.................	4.61	39.16	45.74	10.49		12791		13.24
McAlister, Tobocksey Co..........	2.10	29.71	62.67	5.52		13796		14.27

INDIANA.

Block............................	3.50	32.50	63.00	1.00	0.98	14020		14.5
" "						13588		14.38
Caking...........................						14146		14.64

INDIANA.—*Continued.*

Name or Locality.	Constituents in Per Cent of Total Weight.					Fuel Value per Pound of Coal.		
	Moisture.	Volatile Matter.	Fixed Carbon.	Ash.	Sulphur.	B.T.U. Calculated.	B.T.U. by Calorimeter	Theoretical Evaporation in lbs. from and at 212°.
Cannel.............	3.5	32.5	61.5	2.5		13097		13.56
Clay Co............	5.5	36.0	53.5	5.0		13898		14.39
Daviess Co..........	5.0	36.0	54.5	4.5		13471		13.95
Fountain Co.........	7.0	29.5	63.0	0.05		13621		14.10
Greene Co...........	3.0	41.5	52.0	3.5		13861		14.35
Montgomery Co......	2.0	38.5	57.5	2.0		14146		14.64
Owen Co............	4.0	46.0	46.5	3.5		14481		14.99
Parke Co............	4.0	39.5	55.0	5.5		14062		14.56
Posey Co............	3.5	40.0	55.0	1.5		13660		14.14
Sullivan Co..........	3.5	42.0	48.5	6.0		14362		14.87
Vanderburg Co.......	5.5	44.0	46.0	4.5		13690		14.17
Vermillion Co........	3.5	41.5	49.5	5.5		13661		14.14
Warwick Co.........						13761		14.24

IOWA.

Albia, Monroe Co....	5.16	40.21	45.88	8.75		12997		13.45
Good Cheer..........	10.85	30.32	31.38	27.45	7.32		8702	9.01
Oskaloosa, Marion Co.	5.73	46.54	45.60	2.13		14040		14.53
Ottumwa, Wapello Co.	6.50	41.35	48.25	3.90		13571		14.05

KANSAS.

Cherokee Co....	1.94	36.77	52.45	8.84		13403		13.87

FUELS.

KENTUCKY.

Name or Locality.	Constituents in Per Cent of Total Weight.					Fuel Value per Pound of Coal.		
	Moisture.	Volatile Matter.	Fixed Carbon.	Ash.	Sulphur.	B.T.U. Calculated.	B.T.U. by Calorimeter	Theoretical Evaporation in lbs., from and at 212°.
Caking...............						14391		14.89
Cannel...............						15198		16.76
"						13360		13.84
Clifton, Hopkins Co........		33.32	47.45	16.02	3.21	12054		12.48
Hawesville, Hancock Co.....	3.30	39.00	50.50	7.20	3.373	12490		12.93
Kensee, Whitley Co.........	1.90	32.86	63.10	2.14	0.70	14396		14.90
Lignite..................						9326		9.65
Manchester, Clay Co........	1.20	38.10	60.70	5.80	1.793	14845		15.36
Peach Orchard, Lawrence Co..	3.24	36.56	49.24	11.56	1.793	12875		13.33
MARYLAND.								
Cumberland...............	1.23	15.47	73.51	9.09	0.70	13067		13.53
"						12226		12.65
George's Creek............						13500		13.98
MISSOURI.								
Bevier Mines..............							9890	10.24
Rich Hill, Bates Co.........	2.54	42.62	41.14	13.70		12641		13.08
MONTANA.								
Sand Coulle, Cascade Co.....	3.01	30.23	59.71	7.05		13504		13.98

80　　　　　　　　　STEAM-BOILER PRACTICE.

Name or Locality.	Constituents in Per Cent of Total Weight.					Fuel Value per Pound of Coal.		
	Moisture.	Volatile Matter.	Fixed Carbon.	Ash.	Sulphur.	B.T.U. Calculated.	B.T.U. by Calorimeter	Theoretical Evaporation in lbs. from and at 212°.
NEW MEXICO.								
Ranton, Colfax Co............	3.10	35.00	51.50	10.40	0.61	12966	11756	13.42
Coal.........................	2.35	35.53	50.24	11.88				12.17
OHIO.								
Bellaire, Belmont Co..........	1.53	42.29	47.57	8.61	4.47	13575		14.05
Briar Hill, Mahoning Co......	2.47	31.83	64.25	1.45	0.56	13714		14.2
Brilliant, Jefferson Co........	1.85	37.82	55.62	4.71	1.32	14072		14.57
Brookfield, Noble Co..........	3.27	40.23	42.72	11.78	5.90	12800		13.25
Buchtel, Athens Co...........	5.10	36.97	49.68	8.25	2.41	13037		13.50
Cambridge, Guernsey Co......	5.32	37.46	53.29	3.93	1.38	13687		14.17
Hocking Valley...............	8.25	35.88	53.15	2.72	0.43	13414		13.9
Liberty, Trumbull Co.........	5.91	35.01	55.70	3.38	0.76	11983		12.40
Morgan, Morgan Co...........	4.20	38.65	43.83	13.32	5.37	12403		12.84
Pike, Stark Co................	2.85	39.00	46.69	11.46	3.14	12893		13.34
Salineville, Columbiana Co....	2.32	39.08	52.78	5.82	2.58	13849		14.34
Walnut Creek, Holmes Co.....	4.49	42.50	47.27	5.74	3.41	13594		14.07
OREGON.								
Coos Co.......................	15.45	41.45	34.95	8.05	2.53	11628		12.04
PENNSYLVANIA.								
Anthracite....................						14199		14.70
" "......................						13535		14.01
" "......................						14221		14.72
" , pea..................	2.04	6.36	8.41	13.19		12300		12.73

PENNSYLVANIA.—Continued.

Name or Locality.	Constituents in Per Cent of Total Weight.					Fuel Value per Pound of Coal.		
	Moisture.	Volatile Matter.	Fixed Carbon.	Ash.	Sulphur.	B.T.U. Calculated.	B.T.U. by Calorimeter	Theoretical Evaporation in lbs. from and at 212°.
Anthracite, buckwheat	3.88	3.84	81.32	10.96	0.67	12200		12.63
Antrim, Tioga Co	1.46	21.60	65.12	9.00	2.82	12825		13.27
Bennington, Blair Co	0.98	26.40	65.586	4.78	2.274	14704		15.22
Bernice, Sullivan Co	1.295	8.10	83.344	6.23	1.031	13350		13.82
Cameron, Northumberland Co	1.815	6.18	86.748	4.502	0.755	13557		14.04
Cannel						13143		13.60
Carbon, Huntingdon Co	0.45	16.21	70.601	8.569	4.17	13770		14.25
Centre, Butler Co	2.11	37.57	51.243	7.178	1.894	13372		13.84
Connellsville, Fayette Co	1.26	30.107	59.616	8.233	0.784	13390		13.86
Franklin, Greene Co	1.265	34.685	49.59	13.19	1.27	12309		12.74
Gillesville, McKean Co	0.67	36.065	48.417	13.79	1.058	12653		13.09
Hillsboro, Washington Co	0.77	36.115	48.554	12.415	2.146	12695		13.14
Irwin, Westmoreland Co	1.78	35.36	59.29	2.89	0.68	14232		14.73
Keystone, Somerset Co	1.29	20.865	67.201	8.805	1.839	13012		13.47
Latrobe, Westmoreland Co	1.59	30.945	63.489	3.18	0.796	14135		14.63
Monongahela, Washington Co	1.18	35.83	58.154	4.075	0.761	14130		14.63
Nicholson, Fayette Co	1.06	34.805	53.538	8.165	2.432	13254		13.72
Osceola, Clearfield Co	1.24	24.45	67.045	5.945	1.32	13589		14.07
Perry, Lawrence Co	1.94	39.265	55.828	2.24	0.727	14358		14.86
Pittsburg (av.)	1.80	35.34	54.94	7.92	1.97		13104	13.46
" coking	1.43	30.22	61.87	6.48	1.35		12981	14.9
Reynoldsville	1.20	27.12	65.88	5.80		14415		13.44
Schuylkill Co	2.98	3.38	87.127	5.856	0.657	13153		13.61
Summer Hill, Cambria Co	0.82	19.155	70.175	9.405	0.445	13164		13.62

PENNSYLVANIA.—Continued.

Name or Locality.	Constituents in Per Cent of Total Weight.					Fuel Value per Pound of Coal.		
	Moisture.	Volatile Matter.	Fixed Carbon.	Ash.	Sulphur.	B.T.U. Calculated.	B.T.U. by Calorimeter	Theoretical Evaporation in lbs., from and at 212°.
West Lebanon, Indiana Co........	1.04	36.94	50.85	9.705	1.465	13189	12936	13.65
Youghiogheny....................	1.96	34.06	58.98	5.00			12600	13.39
" "	2.02	32.14	58.96	6.88	0.88			13.03

RHODE ISLAND.

Portsmouth.......................		10.0	84.50	5.50		13829		14.31

TENNESSEE.

Coal Creek, Anderson Co.........		34.20	58.80	6.93	0.79	13936	13167	14.43
Glen Mary, Scott Co.............	2.15	31.47	61.63	4.75	0.94	14168		13.63
Newcombe, Campbell Co..........		33.77	60.64	3.59	1.10	12740		14.67
Soddy, Hamilton Co..............		28.90	56.89	14.21				13.19

TEXAS.

Cedral Mine......................	0.45	21.60	45.70	29.10		9800	9450	10.14
Ft. Worth........................	14.42	30.03	42.53	13.02	1.47		11403	9.78
" "	4.60	34.72	49.27	11.41	1.56			11.80
Lignite..........................						12962		13.41
Sabinas Mine.....................	0.83	29.35	50.18	19.63		11782		12.20

UTAH.

Coalville, Summit Co.............		38.90	56.37	4.30	10.32	12093		13.35
Scofield, Emery Co...............		39.85	47.30	5.55	7.30	13178		13.64

FUELS.

VIRGINIA.

Name or Locality.	Constituents in Per Cent of Total Weight.					Fuel Value per Pound of Coal.		
	Moisture.	Volatile Matter.	Fixed Carbon.	Ash.	Sulphur.	B.T.U. Calculated.	B.T.U. by Calorimeter	Theoretical Evaporation in lbs, from and at 212°.
Carbon Hill	0.40	18.60	71.00	10.00		13192		13.65
Clover Hill	1.339	30.984	56.831	10.132		13103		13.56
Coalbrookdale		24.00	70.80	5.20	0.514	14068		14.50
Pocahontas, Wisconsin Co	6.940	18.832	74.066	5.647	0.761	13718		14.20

WASHINGTON.

Franklin	3.33	33.92	57.68	5.07		13746		14.23
Newcastle	11.70	35.50	46.00	6.04	0.04	12290		12.72
Roslyn, Kittitass Co	2.42	27.37	56.69	13.36	0.16	12475		12.91
Wilkeson, Pierce Co	1.10	35.10	54.50	9.30		13429		13.90

WEST VIRGINIA.

Cedar Grove, Kanawha Co	2.10	34.08	60.67	2.50	0.65	12233		12.66
Elmo, Fayette Co	1.05	23.62	72.67	2.20	0.76	14314		14.81
New River	0.67	26.64	70.66	1.53	0.50	14509		15.02
"						14200		14.70
"						13400		13.87
Pocahontas	0.94	18.19	75.89	4.68	0.30		14273	14.71
"	0.76	18.65	79.26	1.11	0.23		14796	15.50
"	0.84	19.44	75.16	4.03	0.55			
"	0.10	17.00	81.10	1.80				

WYOMING.

Dana, Carbon Co	11.30	42.01	39.69	7.00		12401		12.83
Glenrock	10.20	34.40	51.11	5.29		12865		13.31
Rawlins	7.47	36.05	51.56	4.32		13204		13.66

beneath the earth's surface, whence it is either pumped, or flows to the surface after the manner of operation of an artesian well.

Crude petroleum is dark brown in color, with a perceptible greenish tinge, and has a specific gravity which averages about 0.8. It is composed of a great number of liquid hydrocarbons varying widely in specific gravity and chemical composition, and each separable from the others by fractional distillation. The general composition and heating-power of various petroleum oils are presented in Table No. 32.

Natural Gas.—In its composition natural gas varies greatly. Not only is there a marked difference in composition between the gas from different wells, but also between samples which are taken at different times from the same well. This is very clearly shown in the accompanying Table No. 33 by the results of six analyses made within a period of three months of samples from a well near Pittsburgh, Pa.

Artificial Fuels.—Although artificial fuels serve a useful purpose in steam-making, their use is by no means as extended as that of the natural fuels. The desirability of employing one in preference to the other is dependent largely upon financial considerations; and that fuel is to be chosen which, other things equal, will evaporate the most water for a given total expenditure. Artificial fuels may be broadly classified under the headings charcoal, coke, fuel-gas, and patent fuels.

Charcoal.—Wood, protected from the atmosphere and heated at about 600°, gives up its gaseous or volatile elements, and is converted into charcoal. The best charcoal consists almost entirely of pure carbon; but in so far as the process of manufacture falls short of perfection, so the proportion of carbon is reduced. Charcoal is distinctly the result of a process of carbonization; and under the condition of distillation in vessels externally heated to various temperatures, its quality is improved as its temperature is increased. The results of this process when applied to black alder, previously

FUELS. 85

TABLE No. 32.

COMPOSITION AND CALORIFIC VALUE OF PETROLEUM OILS.

Name of Locality.	Carbon.	Hydrogen	Oxygen.	Nitrogen.
Galicia, West..................	85.3	12.6	2.1	
Hanover, Odessa.......................	80.4	12.7	6.9	
A'sace, Schwabweiler...................	79.5	13.6	6.9	
France, St. Gabian......	86.1	12.7	1.2	
Italy, Neviano de Rossi................. ...	81.9	12.5	5.6	
Roumania, Plojesti.....................	82.6	12.5	4.9	
Russia, Baku...........................	85.3	11.6	3.1	
Canada, Bothwell......................	84.3	13.4	2.3	
" Petrolea...................	84.5	13.5	2.0	
Pennsylvania, Oil Creek................	82.0	14.8	3.2	
"	84.9	13.7	1.4	
Ohio.....................................	84.2	13.1	2.7	
" Mecca.................'........	86.32	13.07	....	0.23
West Virginia.........................	84.3	14.1	1.6	
" "	83.2	13.2	3.6	
" "	83.6	12.9	3.5	
" " Scotia Well	86.62	12.93		
" " Cumberland..	85.2	13.36	....	0.54
California, Hayward Petroleum Co.......	86.93	11.82	1.11	

Dr. H. Gintl gives the following calorific values of American oils :

Petroleum from   B.T.U.
West Virginia................................... 10,180
Pennsylvania................................... 9,963

Since the composition of oils even from the same district varies, their calorific values are presented only as general averages.

86 STEAM-BOILER PRACTICE.

TABLE No. 33.

COMPOSITION AND CALORIFIC VALUE OF NATURAL GAS FROM AMERICAN WELLS.

Name or Locality.	Hydrogen	Marsh-gas	Olefiant Gas.	Carbon Monoxide.	Carbon Dioxide.	Oxygen.	Nitrogen.	Hydrogen Sulphide.	Ethylic Hydride.
INDIANA.									
Anderson..........	1.86	93.07	0.47	0.73	0.26	0.42	3.02	0.15	
Kokomo...........	1.42	94.16	0.30	0.55	0.29	0.30	2.80	0.18	
Marion............	1.20	93.57	0.15	0.60	0.30	0.55	3.42	0.20	
Muncie............	2.35	92.67	0.25	0.45	0.25	0.35	3.53	0.15	
OHIO.									
Findlay...........	1.64	93.35	0.35	0.41	0.25	0.39	3.41	0.20	
Fostoria...........	1.89	92.84	0.20	0.55	0.20	0.35	3.82	0.15	
St. Mary's........	1.94	93.85	0.20	0.44	0.23	0.35	2.98	0.21	
PENNSYLVANIA.									
Results of six analyses made within period of three months of samples from well near Pittsburg.	9.64	57.85	0.80	1.00	0.00	2.10	23.41	......	5.20
	14.45	75.16	0.60	0.30	0.30	1.20	2.89	......	4.80
	20.02	72.18	0.70	1.00	0.80	1.10	0.00	......	3.60
	26.16	65.25	0.80	0.80	0.60	0.80	0.00	......	5.50
	29.03	60.70	0.98	0.58	0.00	0.78	0.00	......	7.92
	35.92	49.58	0.60	0.40	0.40	0.80	0.00	......	12.30

The calorific value of natural gas of average composition may be taken at about 1000 B.T.U.

dried at about 300°, as reported by Violette, are presented in Table No. 34.

TABLE No. 34.

COMPOSITION OF CARBON PRODUCED AT VARIOUS TEMPERATURES.

Temperature of Carbonization.	Constituents of the Solid Product.				
	Carbon.	Hydrogen.	Oxygen.	Nitrogen and Loss.	Ash.
302° F.	47.51	6.12	46.29	0.08	47.51
392	51.82	3.99	43.98	0.23	39.88
482	65.59	4.81	28.97	0.63	32.98
572	73.24	4.25	21.96	0.57	24.61
662	76.64	4.14	18.44	0.61	22.42
810	81.64	4.96	15.24	1.61	15.40
1873	81.97	2.30	14.15	1.60	15.30

Peat-charcoal, produced by the carbonization of ordinary air-dried peat, is very friable and porous, and extremely difficult to handle without reducing it to very small particles almost powdery in their character. Although it is easily ignited and burns readily, its physical characteristics are such as to prevent its general use.

Coke.—The residual product of the carbonization of bituminous coal is known as coke. By this process the hydrocarbon gases are expelled, and the coal is reduced to a substance somewhat porous in its character and consisting almost entirely of carbon. The coke produced by the partial combustion of coal in coke-ovens, is dark gray in color, hard, porous, and brittle, with a slightly metallic lustre. That resulting as a by-product from the distillation of gas in the retorts of gas-works is not so hard, ignites more readily, and burns with a draft less intense than that required for the combustion of coke which has been formed by the first method. It is, therefore, better adapted as a fuel for steam-boiler furnaces.

The quality of such coke is affected by the temperature and by the pressure under which distillation takes place. From experiments by Steavenson, reported to the Iron and

Steel Institute at Newcastle, Eng., it appeared that with a furnace of special construction and coal of the following composition,—

Oxygen	6.7	per cent
Carbon	84.9	"
Hydrogen	4.5	"
Nitrogen	1.0	"
Sulphur	0.6	"
Ash	2.3	"

the yield was about 60 per cent of coke of the following composition:

Carbon	96.2	per cent
Ash	3.8	"

Hence the composition and relative weight of the materials lost in coking were—

Carbon	68.1	per cent
Hydrogen	11.2	"
Nitrogen	2.5	"
Sulphur	1.6	"
Oxygen	16.6	"

The average heating-power of coke ranges from about 13,000 B. T. U. to nearly 15,000 B. T. U.

Fuel-gas.—Although carbonic oxide and hydrogen are combustibles, the production of which is incident to the combustion of all fuels, they are never independently manufactured for use as fuels. But hydrogen and carbon, associated in the form of volatile hydrocarbons, serve most excellently the purposes of fuels, although their cost must determine their efficiency. Notwithstanding the fact that illuminating-gas made from bituminous coal by distillation in retorts has been in common use for nearly a century, the idea of directly converting a solid fuel into one of gaseous form for its readier utilization for producing heat has only been carried into practice during the past twenty-five or thirty years. Gas as a fuel first appeared in the form of "producer" gas, and was primarily introduced for metallurgical purposes. In its manufacture, air, mixed with water-vapor, was passed, under powerful pressure, through a thick bed of burning coal. As a result the coal was only burned to carbonic oxide, while the watery vapor was decomposed so that the resulting gas from the producer was a mixture of about one half nitrogen and one fourth carbonic oxide, with varying proportions of hydrogen and hydrocarbons. Its composition depends upon the proportions of the elements in the original fuel. This process, however, inherently consumes about one third of the total calorific value of the fuel, thereby reducing by that amount the resultant heating-power.

Water-gas is the result of a somewhat similar process, which differs principally from that employed in the manufacture of producer-gas in that it is intermittent, first air and then steam being forced through a bed of incandescent fuel. For illuminating purposes this gas is carburetted, so that it actually exceeds by volume the value of coal-gas.

Including natural gas the relative volumes and weights of gaseous fuels are:

	By Weight.	By Volume.
Natural gas	1,000	1,000
Coal-gas	949	666
Water-gas	292	292
Producer-gas	76.5	130

By weight and volume the composition of these gases is given in Table No. 35.

In this table the natural gas was from Findlay, Ohio, the coal-gas was probably an average sample purified for illuminating purposes, the water-gas was made for heating and consequently unpurified, and the producer-gas was made from anthracite at the Pennsylvania Steel Works.

TABLE No. 35.

COMPOSITION OF FUEL-GASES.

Constituents.	By Volume.				By Weight.			
	Natural Gas.	Coal-gas.	Water-gas.	Producer-gas.	Natural Gas.	Coal-gas.	Water-gas.	Producer-gas.
Hydrogen	2.18	46.0	45.0	6.0	0.268	8.21	5.431	0.458
Marsh gas	92.60	40.0	2.0	3.0	90.383	57.20	1.931	1.831
Carbonic oxide	0.50	6.0	45.0	23.5	0.857	15.02	76.041	25.095
Olefiant gas	0.31	4.0	0.0	0.0	0.531	10.01	0.000	0.000
Carbonic acid	0.26	0.5	4.0	1.5	0.700	1.97	10.622	2.517
Nitrogen	3.61	1.5	2.0	65.0	6.178	3.75	3.380	69.413
Oxygen	0.34	0.5	0.5	0.0	0.666	1.43	0.965	0.000
Water vapor	0.00	1.5	1.5	1.0	0.000	2.41	1.630	0.686
Sulphydric acid	0.20				0.417			

Patent Fuels.—Under this title may be classed a large variety of prepared fuels, consisting in the main of the particles of some finely divided combustible pressed and cemented together by a substance possessing the necessary adhesive and inflammable properties.

In the process of coal-mining, sorting, and shipping, a considerable amount is broken into fragments too small for ordinary commercial use. This refuse, commonly denominated "culm," possesses practically the same calorific value as the coal of which it originally formed a part; but its finely divided character is not conducive to its successful use in an ordinary boiler-furnace. It may, however, by special machinery, be mixed with sufficient pitch or coal-tar, and moulded into lumps of desirable size.

In this country the relative price of coal is so low, as compared with the cost of manufacture of such pressed fuel, that the financial return hardly warrants the attempt to thus utilize the culm. In France, and some other European countries, fuel of this character, in the form of "briquettes," is regularly made of coal-dust,—bituminous and semi-anthracite,—and quite extensively used. To some extent the slow progress made in the manufacture of briquettes in this country is doubtless due to the imperfect systems of washing and jigging which are necessary to reduce the percentage of ash, which never ought to exceed 10 per cent in such fuel.

The attempt has also been made, with varying success, according to the conditions, to feed the coal in the form of dust directly to the boilers, by forcing it into a strong air-current, which thus spreads it throughout the furnace, while at the same time furnishing the oxygen necessary for combustion.

By means of glue, tar, pitch, resin, and the like, sawdust, charcoal, peat, tan, and similar refuse have been cemented together for use as a fuel. But, except in comparatively few instances, the cost of manufacture of prepared fuels has, in this country at least, rendered them but little, if any, more economical than coal.

CHAPTER V.

EFFICIENCY OF FUELS.

Measure of Efficiency.—The ultimate efficiency of a fuel should be expressed in the total amount of heat it is capable of generating, or, in other words, by its calorific power. The proportion of that heat which is utilized depends upon the efficiency of the boiler or other heat-abstracting device. Commercially, however, the heat-values of fuels are generally measured relatively to each other, and expressed in the number of pounds of water evaporated per pound of fuel. In practice, the physical character of the fuel, the form and construction of the boiler and furnace, the amount of air supplied, and other conditions, have an important influence upon the attainable result. In fact, the effect of these variables is such as to render an accurate comparison of fuels a difficult matter.

In their ultimate efficiency the calorific power of fuels may very properly be considered relatively to an established standard. As carbon is the most important element in the composition of all fuels, it may reasonably be selected as such a standard. For the purposes of comparison, Table No. 36 has been prepared to show the efficiency of fuels as measured in thermal units, determined by analysis or calorimetric test when compared with pure carbon as a standard, having a thermal value of 14,650 B. T. U. Measured by this standard, a coal having very little ash and a large amount of hydrogen may, because of its extremely high heating-power, actually show an efficiency above 100 per cent when compared with carbon. As the theoretical efficiency of a fuel can never be realized in practice, there have been incorporated in this table the fuel-efficiency and the corresponding number of pounds of water evaporated from and at 212° per pound of fuel at various boiler-efficiencies.

TABLE No. 36.

EFFICIENCY OF FUELS.

British Thermal Units in One Pound of Fuel.	Water Evaporated from and at 212° per Pound of Fuel.	Theoretical Efficiency of Fuel as compared with Pure Carbon.	Efficiency of Boiler.									
			90 per cent.		80 per cent.		70 per cent.		60 per cent.		50 per cent.	
			Water Evaporated from and at 212° per Pound of Fuel.	Efficiency of Fuel.	Water Evaporated from and at 212° per Pound of Fuel.	Efficiency of Fuel.	Water Evaporated from and at 212° per Pound of Fuel.	Efficiency of Fuel.	Water Evaporated from and at 212° per Pound of Fuel.	Efficiency of Fuel.	Water Evaporated from and at 212° per Pound of Fuel.	Efficiency of Fuel.
14650	15.2	100.0	13.6	90.0	12.1	80.0	10.6	70.0	9.1	60.0	7.6	50.0
14500	15.0	99.0	13.5	89.0	12.0	79.2	10.5	69.3	9.0	59.4	7.5	49.5
14250	14.8	97.3	13.3	87.6	11.8	77.8	10.3	68.1	8.9	58.4	7.4	48.7
14000	14.5	95.6	13.0	86.0	11.6	76.5	10.1	66.9	8.7	57.4	7.3	47.8
13750	14.2	93.9	12.8	84.5	11.4	75.1	10.0	65.7	8.6	56.3	7.1	47.0
13500	14.0	92.2	12.6	83.0	11.2	73.8	9.8	64.5	8.4	55.3	7.0	46.1
13250	13.7	90.5	12.3	81.5	11.0	72.4	9.6	63.4	8.3	54.3	6.9	45.3
13000	13.5	88.8	12.1	79.9	10.8	71.0	9.4	62.2	8.1	53.3	6.7	44.4
12750	13.2	87.1	11.9	78.4	10.5	69.7	9.2	61.0	7.9	52.3	6.6	48.6
12500	12.9	85.4	11.6	76.9	10.3	68.3	9.0	59.8	7.7	51.2	6.5	42.7
12250	12.7	83.7	11.4	75.3	10.1	67.0	8.9	58.6	7.6	50.2	6.4	41.9
12000	12.4	82.0	11.2	73.8	9.9	65.6	8.7	57.4	7.5	49.2	6.2	41.0
11750	12.2	80.3	11.0	72.3	9.7	64.2	8.5	56.2	7.3	48.2	6.1	40.2
11500	11.9	78.6	10.8	70.7	9.6	62.9	8.4	55.0	7.1	47.2	6.0	39.3
11250	11.7	76.9	10.5	69.2	9.3	61.5	8.2	53.8	7.0	46.1	5.9	38.5
11000	11.4	75.2	10.2	67.7	9.1	60.2	8.0	52.6	6.8	45.1	5.7	37.6
10750	11.1	73.5	10.0	66.2	8.9	58.8	7.8	51.5	6.7	44.1	5.6	36.8
10500	10.9	71.8	9.8	64.6	8.7	57.4	7.6	50.3	6.5	43.1	5.4	35.9
10250	10.6	70.0	9.5	63.0	8.5	56.0	7.4	49.1	6.4	42.0	5.3	35.0
10000	10.4	68.3	9.3	61.5	8.3	54.7	7.2	47.8	6.3	41.0	5.2	34.2
9750	10.1	66.6	9.1	59.9	8.1	53.3	7.1	46.6	6.1	40.0	5.1	33.3
9500	9.8	64.9	8.8	58.4	7.9	51.9	6.9	45.4	5.9	38.9	4.9	32.5
9250	9.6	63.2	8.6	56.9	7.7	50.6	6.7	44.2	5.7	37.9	4.8	31.6
9000	9.3	61.4	8.4	55.3	7.5	49.1	6.5	43.0	5.6	36.8	4.7	30.7
8750	9.1	59.7	8.2	53.7	7.3	47.8	6.3	41.8	5.4	35.8	4.6	29.8
8500	8.8	58.0	7.9	52.2	7.0	46.4	6.2	40.6	5.3	34.8	4.4	29.0

Evidently, the actual efficiency of a given fuel is here, as in all cases, dependent upon the efficiency of the boiler. It is obvious, however, that an evaporation of 15.2 pounds of water from and at 212° per pound of best coal represents an ideally perfect result, with 100 per cent efficiency of both fuel and boiler, unless the fuel contains sufficient volatile matter to raise its total heat above 14,650 B. T. U.

Unit of Evaporation.—For the purposes of practical comparison there is usually determined the number of pounds of water of a given temperature that can be evaporated into steam of a given pressure by the combustion of one pound of the fuel. This is reduced for direct comparison to the standard of temperature of water at 212°, and steam of atmospheric pressure; namely, of a temperature of 212°. Under these conditions, as no heat is expended in heating the water, the amount of heat required to evaporate one pound of water is equal to the latent heat of steam at atmospheric pressure; that is, 965.7 B. T. U. The evaporation per pound of coal or combustible, as reduced to this basis, is known as the *unit of evaporation*.

In illustration of the method of calculation, suppose that a given test indicates that 8.73 pounds of water fed to the boiler at 120° have been evaporated into steam of 83.3 pounds gauge-pressure by the combustion of one pound of the fuel under test, without correction for moisture in steam and fuel. The absolute steam-pressure is $83.3 + 14.7 = 98$ pounds, and, by Table No. 7, the total amount of heat contained in one pound of steam of this pressure is 1213.40 B. T. U.; while, by Table No. 5, the total heat of the water of 120° temperature is 120.149 B. T. U. Evidently, then, the amount of heat which was imparted to one pound of water at 120° in order to convert it into steam of 83.3 pounds gauge-pressure was $1213.40 - 120.149 = 1093.251$ B. T. U. As the latent heat of steam at atmospheric pressure is 965.7 B. T. U., the evaporation of 8.73 pounds of water under the stated conditions is equivalent to the evaporation of

$$\frac{1093.251}{965.7} \times 8.73 = 9.87 \text{ pounds,}$$

from and at 212°.

For ultimate comparison of fuels the results should be corrected for moisture in the steam and in the fuel. If, under the conditions of the test just used for illustration, the steam had contained 1.2 per cent and the coal 3.5 per cent of moisture, the method of correction would be as follows:

The actual proportion of dry steam would be $100 - 1.2 = 98.8$ per cent, and that of dry coal $100 - 3.5 = 96.5$ per cent. The amount of water evaporated into dry steam from and at $212°$ per pound of fuel would, therefore, be $9.87 \times 0.988 = 9.75$ pounds; and the evaporation of dry steam per pound of dry fuel would be $9.75 \div 0.965 = 10.10$ pounds; or, combined in one calculation,

$$\frac{9.87 \times 0.988}{0.965} = 10.10 \text{ pounds.}$$

To ascertain the equivalent evaporation per pound of combustible in the fuel, the proportion of ash must be ascertained by careful weighing and deducted from the total fuel burned. If the ash in the fuel from which the preceding results were obtained had amounted to 6.4 per cent, the evaporation of water from and at $212°$ into dry steam would have been

$$\frac{10.10}{1.00 - 0.064} = 10.79 \text{ pounds.}$$

Table No. 37, calculated by the method previously explained, presents a series of factors by means of any one of which, corresponding to the given temperature of feed-water and pressure of steam, the evaporative result obtained may be transformed into the equivalent evaporation from and at $212°$. Thus, taking the conditions of 83.3 pounds and $120°$ temperature already given, the factor (ascertained by interpolation) is 1.132, which, multiplied by 8.73, gives 9.88 * pounds of water evaporated from and at $212°$. Evidently, this table may be used in a converse manner to determine what conditions of feed-temperature and steam-pressure may be equivalent to a stated evaporation from and at $212°$.

---

* The slight difference between this and the calculated result is due to the fact that the numbers in the table are not carried out to more decimal places.

Thus, for instance, an evaporation of 9.87 pounds from and at 212° is equivalent to 9.87 ÷ 1.149 = 8.59 pounds from water at 100° into steam at 70 pounds gauge-pressure.

TABLE No. 37.

FACTORS OF EVAPORATION.

Temperature of Feed-water, in Degrees Fahr.	Pressure in Pounds per Square Inch above the Atmosphere.									
	0	20	40	50	60	70	80	90	100	120	140	160
32°	1.187	1.201	1.211	1.214	1.217	1.219	1.222	1.224	1.227	1.231	1.234	1.237
35	1.184	1.198	1.208	1.211	1.214	1.216	1.219	1.221	1.224	1.228	1.231	1.234
40	1.179	1.193	1.203	1.206	1.209	1.211	1.214	1.216	1.219	1.223	1.226	1.229
45	1.173	1.187	1.197	1.200	1.203	1.205	1.208	1.210	1.213	1.217	1.220	1.223
50	1.168	1.182	1.192	1.195	1.198	1.200	1.203	1.205	1.208	1.212	1.215	1.218
55	1.163	1.177	1.187	1.190	1.193	1.195	1.198	1.200	1.203	1.207	1.210	1.213
60	1.158	1.172	1.182	1.185	1.188	1.190	1.193	1.195	1.198	1.202	1.205	1.208
65	1.153	1.167	1.177	1.180	1.183	1.185	1.188	1.190	1.193	1.197	1.200	1.203
70	1.148	1.162	1.172	1.175	1.178	1.180	1.183	1.185	1.188	1.192	1.195	1.198
75	1.143	1.157	1.167	1.170	1.173	1.175	1.178	1.180	1.183	1.187	1.190	1.193
80	1.137	1.151	1.161	1.164	1.167	1.169	1.172	1.174	1.177	1.181	1.184	1.187
85	1.132	1.146	1.156	1.159	1.162	1.164	1.167	1.169	1.172	1.176	1.179	1.182
90	1.127	1.141	1.151	1.154	1.157	1.159	1.162	1.164	1.167	1.171	1.174	1.177
95	1.122	1.136	1.146	1.149	1.152	1.154	1.157	1.159	1.162	1.166	1.169	1.172
100	1.117	1.131	1.141	1.144	1.147	1.149	1.152	1.154	1.157	1.161	1.164	1.167
105	1.111	1.125	1.135	1.139	1.141	1.143	1.146	1.148	1.151	1.155	1.158	1.161
110	1.106	1.120	1.130	1.133	1.136	1.138	1.141	1.143	1.146	1.150	1.153	1.156
115	1.101	1.115	1.125	1.128	1.131	1.133	1.136	1.138	1.141	1.145	1.148	1.151
120	1.096	1.110	1.120	1.123	1.126	1.128	1.131	1.133	1.136	1.140	1.143	1.146
125	1.091	1.105	1.115	1.118	1.121	1.123	1.126	1.128	1.131	1.135	1.138	1.141
130	1.085	1.099	1.109	1.112	1.115	1.117	1.120	1.122	1.125	1.129	1.132	1.135
135	1.080	1.094	1.104	1.107	1.110	1.112	1.115	1.117	1.120	1.124	1.127	1.130
140	1.075	1.089	1.099	1.102	1.105	1.107	1.110	1.112	1.115	1.119	1.122	1.125
145	1.070	1.084	1.094	1.097	1.100	1.102	1.105	1.107	1.110	1.114	1.117	1.120
150	1.065	1.079	1.089	1.092	1.095	1.097	1.100	1.102	1.105	1.109	1.112	1.115
155	1.059	1.073	1.083	1.086	1.089	1.091	1.094	1.096	1.099	1.103	1.106	1.109
160	1.054	1.068	1.078	1.081	1.084	1.086	1.089	1.091	1.094	1.098	1.101	1.104
165	1.049	1.063	1.073	1.076	1.079	1.081	1.084	1.086	1.089	1.093	1.096	1.099
170	1.044	1.058	1.068	1.071	1.074	1.076	1.079	1.081	1.084	1.088	1.091	1.094
175	1.039	1.053	1.063	1.066	1.069	1.071	1.074	1.076	1.079	1.083	1.086	1.089
180	1.033	1.047	1.057	1.060	1.063	1.065	2.068	1.070	1.073	1.077	1.080	1.083
185	1.028	1.042	1.052	1.055	1.058	1.060	1.063	1.065	1.068	1.072	1.075	1.078
190	1.023	1.037	1.047	1.050	1.053	1.055	1.058	1.060	1.063	1.067	1.070	1.073
195	1.018	1.032	1.042	1.045	1.048	1.050	1.053	1.055	1.058	1.062	1.065	1.068
200	1.013	1.027	1.037	1.040	1.043	1.045	1.048	1.050	1.053	1.057	1.060	1.063
205	1.008	1.022	1.032	1.035	1.038	1.040	1.043	1.045	1.048	1.052	1.055	1.058
210	1.004	1.017	1.027	1.030	1.033	1.035	1.038	1.040	1.043	1.047	1.050	1.053
212	1.002	1.000	.....	.....	.....	.....	.....	.....	.....	.....	.....	.....

Relative Efficiency of Various Coals.—Although the preceding applies to all classes of fuel, the greatest interest centres in the practical calorific value of various kinds of coal; for upon this fuel, above all others, is general steam-boiler practice most dependent. It must already have become evident that the apparent efficiency of the coal and of the boiler in connection with which it is burned are interdependent. Increased calorific value on the part of the coal insures an increase in the output of the boiler; while an improvement in the proportions of the boiler, its appurtenances or its method of operation, whereby its steaming-power is increased per pound of coal, likewise raises the practical efficiency of the coal.

In general it may be stated that any furnace is well adapted to the combustion of anthracite and semi-bituminous coals containing less than 20 per cent of volatile matter. For coals containing between 20 and 40 per cent a plain grate-bar furnace with fire-brick arch thrown over it is desirable, because of its ability to keep the furnace-chamber hot. For coals which contain over 40 per cent of volatile matter a furnace is desirable which is surrounded by fire-brick, with a large combustion-chamber and special appliances for introducing very hot air to the gases distilled from the coal. A separate gas-producer and combustion-chamber, arranged for heating both air and gases before they unite, serves the same purpose.

The practical efficiency of a given coal is dependent not only on its chemical composition and theoretical heat-value, but to a great degree upon the percentage of ash and moisture which it contains, and upon the size of the respective pieces or particles, both absolutely and relatively to each other.

An efficiency of 100 per cent on the part of either the coal or the boiler is an absolute impossibility because of certain losses which are incident to the combustion of the coal and the operation of the boiler. Of these losses some are inevitable, while others may be diminished or avoided.

The unavoidable losses are:

First. The heat lost by converting into steam the water contained in the coal, and in the air used in burning it, as well as that formed by the burning of the hydrogen and the heating of the steam thus formed to the temperature at which the gases leave the stack.

Second. The heat necessary to raise to the stack temperature the carbonic acid gas formed by burning the carbon, the nitrogen originally present in the air from which the oxygen has been taken to form carbonic acid, the sulphurous acid, and the excess of air which is supplied to secure perfect combustion. There is also a loss through heat in the hot ashes removed from the ash-pit, as well as through the unconsumed carbon remaining in the ash. A further loss occurs through radiation from the boilers and walls, which can, by careful construction and covering, be greatly reduced but never eliminated.

The losses which are more or less avoidable are:

First. Those due to incomplete combustion, as evidenced in the presence of smoke and carbonic oxide in the flue-gases and in unconsumed coal in the ashes. This latter loss is due to the original small size or the subsequent decrepitation of the coal, which results in the dropping of more or less of it through the grates without being consumed. In addition a small amount of hydrogen or marsh-gas may pass out with the gases.

Second. Loss from excess of air, due to the fact that to secure practically perfect combustion air must be supplied considerably in excess of the theoretical quantity chemically required for combustion. This loss is twofold, being dependent upon the quantity of unused oxygen and associated nitrogen and upon the moisture in the air.

Third. The loss resulting from too high temperature of the gases leaving the boiler. This loss, except in so far as it is influenced by the air-supply and the rate of combustion, is dependent upon the design of the boiler and its appurtenances, and therefore is not chargeable to the character of the fuel. It is one of the most important factors in fuel-efficiency.

Fourth. Loss of heat by removing ashes at too high a temperature. This, by care, may be reduced but not entirely avoided.

Fifth. Loss by radiation. This may be reduced by increasing the thickness of walls and covering all exposed portions of the boiler. But from a practical standpoint it can never be entirely avoided.

The influence of these various sources of loss upon the efficiency of fuels and boilers will be considered in succeeding pages. Independent of such consideration the relative efficiency of various coals, as indicated by comparative tests, may, however, be here introduced. It is already evident that, for the purposes of strict comparison of evaporative powers, coals should be tested under identical conditions. What is more, all ordinary grades of coal should be tested under such a variety of boilers and rates of combustion, air-supply, and draft that they may be intelligently compared one with the other under all ordinary conditions. Such extensive and strictly comparable results do not, as yet, exist, although already much careful work has been and is now being done to furnish such a basis of comparison.

At the present time it is only possible to compare with each other the results in certain groups of tests, but only to a limited extent to correspondingly compare the results in one group with those in another. By such comparisons as are allowable, it is possible to approximate with reasonable accuracy to the relative values of the coals under consideration.

A series of such results, compiled from the reports of numerous tests by Barrus,* is presented in Table No. 38. In

TABLE No. 38.

COMPARATIVE EVAPORATIVE EFFICIENCY OF VARIOUS COALS.

Kind of Coal.	Water Evaporated from and at 212° by One Pound of Dry Coal.	Relative Efficiency in per cent. Cumberland = 100.
Cumberland	11.04	100
Anthracite, broken.............................	9.79	89
Anthracite, chestnut	9.40	85
Two parts pea and dust and one part Cumberland..	9.38	85
Two parts pea and dust and one part culm	9.01	82
Anthracite, pea	8.86	80
Nova Scotia culm	8.42	76

* "Boiler-tests." George H. Barrus.

harmony with the results obtained in a large number of tests, the evaporation per pound of combustible in anthracite broken coal has been taken, as a standard of comparison, in round numbers at 11.0 pounds from and at 212°. With an average of 11 per cent of ash this is equivalent to 9.79 pounds of water evaporated from and at 212° by one pound of coal.

Influence of Ash.—Beyond its indication of the relative presence of incombustible matter, the influence of ash upon the thermal efficiency of coal is threefold. Its presence is the measure of the loss through the amount of unconsumed carbon which it contains, through the heat lost when the ash is removed in a heated condition, and through the influence of the ash in clogging the fire and preventing free combustion. From a commercial standpoint its presence, furthermore, proportionately increases the original cost of freight and handling per heat-unit derived, as well as the subsequent expense incident to its own removal and transportation to a proper dumping-place,—a fact which demands careful consideration.

The percentage of carbon, either in the form of cinder or decrepitated coal, which eventually forms a part of the ash, can be somewhat reduced by skilful manipulation of the fire. But as the greatest saving of such carbon lies in slower and gentler firing and in admitting more air, the efficiency, as a whole, is liable to be lowered rather than raised if the attempt to economize is carried to an extreme. The carbon ordinarily present in ash varies greatly, but may be broadly stated to range between 10 and 60 per cent. The loss by carbon in the ash is, therefore, dependent upon the percentage of ash in the coal, as is shown in Table No. 39.

It is obvious that the greater the amount of hot ashes resulting from the combustion of a given amount of coal, the greater will be the loss of heat when these ashes are removed. As a means of clogging the grates, preventing free combustion and necessitating extra work on the part of the fireman, with a resulting excess of air while the fire-doors are open, a

EFFICIENCY OF FUELS.

TABLE No. 39.
LOSS OF FIXED CARBON ON ACCOUNT OF CARBON IN ASH.

Per cent of Carbon in Ash.	Per cent of Ash in Coal Fired.									
	9%	10%	11%	12%	13%	14%	15%	16%	17%	18%
75	31.76	35.70	39.73	43.88	48.14	52.47	56.94	61.51	66.23	71.05
70	24.70	27.75	30.89	34.11	37.43	40.80	44.27	47.83	51.51	55.26
65	19.65	22.09	24.59	27.16	29.79	32.48	35.23	37.07	41.00	43.98
60	15.87	17.84	19.86	21.93	24.07	26.23	28.47	30.76	33.11	35.52
55	12.94	14.54	16.18	17.87	19.60	21.37	23.19	24.05	26.98	28.94
50	10.59	11.90	13.24	14.63	16.04	17.49	18.98	20.50	22.08	23.68
45	8.66	9.76	10.83	11.97	13.12	14.31	15.52	16.77	18.06	19.38
40	7.05	7.92	8.82	9.74	10.69	11.66	12.64	13.66	14.72	15.79
35	5.69	6.40	7.12	7.87	8.63	9.40	10.21	11.04	11.88	12.75
30	4.53	5.10	5.66	6.27	6.87	7.43	8.12	8.78	9.45	10.14
25	3.52	3.96	4.41	4.96	5.34	5.82	6.31	6.83	7.35	7.89
20	2.64	2.97	3.30	3.65	4.02	4.36	4.74	5.13	5.52	5.92
15	1.86	2.09	2.33	2.57	2.82	3.09	3.35	3.61	3.89	4.18
10	1.17	1.32	1.47	1.62	1.77	1.93	2.10	2.27	2.45	2.62
5	0.56	0.63	0.69	0.76	0.84	0.91	0.99	1.07	1.15	1.24
1	0.10	0.12	0.13	0.14	0.16	0.17	0.18	0.20	0.22	0.24

large amount of ash in the fuel exerts a very important influence.

Influence of Size of Coal.—A still further, and under certain conditions a very important, influence upon the efficiency of coals, particularly the anthracites, is exerted by the size of their respective pieces or particles. For the freest burning they should be as nearly of a size as possible; hence the screening process at the mines, and their sale in stated sizes. In the smallest sizes of anthracite, consisting of culm and screenings, or slack, the inherent dust and minute particles render them difficult coals to burn unless mixed with a certain proportion of bituminous coal and burned upon special grates, with an intensity of draft which can only be economically produced by mechanical means. This feature becomes more pronounced as the coal becomes finer, and makes legitimate comparison with other coals a somewhat difficult matter. But the fact that cheap fuels are generally the finest, and as increased commercial efficiency can usually be secured by their utilization, the best methods of burning them are of deep interest.

Broadly stated, the requisites to success in the combustion of small anthracites are,* "first, draft; and, second, manipulation of the fires. Of course no fire will burn without draft, and the greater the amount of fire in a given space the stronger the draft must be to properly consume the fuel. With the larger sizes of fuel this question of draft is less prominent, but when we come to burn the smaller sizes the draft is of the utmost importance. The coal will pack on the grate, and, owing to the way the pieces will arrange themselves close together, it will be impossible to get sufficient air through the bed of fire unless the draft is strong enough to displace the smaller particles. . . . Another and perhaps a better reason is that the proportion of ash is somewhat greater with the smaller coals, and as the fire is but a thin crust on top of this, it follows that a somewhat stronger draft will be required to get the necessary volume of air through the bed of ashes and the closely packed crust of the fire on top."

The matter of burning the smaller sizes of refuse coal will be still further considered from a commercial standpoint, but the conditions which control the successful combustion of such fuel may be here discussed. While the terms "screenings" and "slack" are generally applied to the refuse of local coal-yards, the term "culm" is distinctly restricted in its application to the refuse or waste from the coal-mines. Originally applied to any mixture below pea coal, this term has become more restricted in its meaning, as the smaller sizes have been removed in the ordinary process of screening, until at the present time it is in some localities applied to that which cannot be sold as buckwheat, but which has had a part of the dust washed out. It is, therefore, evident that in any dis-

---

* "The Economy of Small Anthracite Coals." Geo. H. Ward. A paper read before the Engineers' Club of Brooklyn, N. Y. 1892.

cussion of the use of a refuse fuel more than its name is necessary to determine its exact character.

Although there is considerable variety in the size classification of coal at different mines, the generally accepted dimensions, as determined by the limiting diameters of the perforations in the screens, are about as presented in Table No. 40.

TABLE No. 40.

SIZES OF COAL.

Designation.	Diameter of Perforation over which Coal will pass.	Diameter of Perforation through which Coal will pass.
Dust...............................	....	3/32 inch.
No. 3 Buckwheat...................	3/32 inch	3/16 "
Bird's-eye.........................	1/8 "	5/16 "
No. 2 Buckwheat, or Rice...........	3/16 "	3/8 "
No. 1 Buckwheat...................	3/8 "	9/16 "
Pea................................	9/16 "	7/8 "
Chestnut..........................	5/8 "	$1\frac{1}{2}$ "
Small stove.......................	1 "	$1\frac{3}{4}$ "
Large stove.......................	$1\frac{1}{4}$ "	$2\frac{1}{4}$ "
Egg...............................	$2\frac{1}{4}$ "	$2\frac{3}{4}$ "
Broken............................	$2\frac{3}{4}$ "	4 "
Steamboat.........................	4 "	7 "

The important factors in the successful burning of small anthracite coals are:

First. Intense draft, preferably mechanically produced and applied beneath the grate.

Second. Large grate-area.

Third. Grate constructed with a practically plain surface to prevent lodgment of coal.

Fourth. Grate of proper design for ready removal of ash and clinker.

Fifth. Air-spaces not over $\frac{1}{16}$ to $\frac{3}{16}$ inch wide, except for No. 1 buckwheat or bituminous slack, for which they may be $\frac{1}{4}$ inch wide.

Sixth. Arrangements to allow of the feeding and cleaning of fires without excessive opening of doors or dropping of ashes into ash-pit.

Seventh. Thin fires and frequent and careful firing. The thickness of the bed should diminish with the rate of combustion.

Eighth. Reduction of draft above the fire as the rate of combustion decreases. The fine character of the material, as a result of which its particles pack closely together with but small interstitial spaces, makes strong draft imperative in order to secure the passage of the proper amount of air through the bed of fuel. Owing to the usually somewhat complicated grate or feeding arrangements for the burning of small anthracites, forced draft is generally applied beneath the grates, but may under certain conditions be favorably assisted by supplementary induced draft.

It appears that the combustion of small anthracites is more perfect when the coal remains undisturbed, or as nearly as possible in the condition in which it was put upon the fire, instead of being turned over so that the partially consumed and unconsumed coal are mixed together. For such fuel the travelling-grate is particularly adapted, as it leaves the fuel undisturbed, and makes possible a gradation of the draft to meet the varying conditions incident to the progress of the grate and the combustion of the fuel upon it.

Whatever the size of these smaller grades of fuel, certain special furnace arrangements are necessary. The smaller they are the more extensive and expensive the appliances; and for each special arrangements are necessary. It is, therefore, undesirable and positively uneconomical to mix the sizes. Thus a mixture of dust with pea or chestnut coal, while burning more freely because of the easier passage of air through it, will not give as good evaporative results as an intermediate size containing less of the larger and the smaller pieces or dust; that is, having all of its pieces nearer a size. This difference is due to the fact that in the mixture the fine coal is completely consumed before the larger pieces.

The advantage of mixing a slight amount of bituminous coal with the smaller anthracites is well known. The somewhat glutinous character of the former, when burning, serves to make a more coherent mass of the entire body of fuel, thereby preventing the dust from being blown through the flues or dropped through the grates, while keeping the bed more open and thereby increasing the rate of combustion. Culm and rice coal thus fired give results in total evaporation which cannot be reached by the same anthracites alone.

The experiments of Coxe have demonstrated that "the temperature developed by the burning of the smaller coals decreases with the size of the coal; this naturally involves a larger heating-surface in the boiler in order to develop the same number of horse-powers; that is to say, if you are burning pea coal, and obtaining one horse-power for every nine square feet of heating-surface, you would probably require from 20 to 25 per cent more heating-surface if you are using No. 3 buckwheat; although you may be evaporating practically the same amount of water per pound of coal."

Table No. 41 gives results of tests * of small anthracites by Coxe.

Tests 1 to 4 inclusive were made upon two Stirling water-tube boilers, each having 1725 square feet of heating-surface and 55 square feet of grate; while No. 5 was made upon a cylinder-boiler with drums and connecting tubes, the total heating-surface being 1862 square feet and 68.75 square feet of grate. All the boilers were equipped with Coxe travelling-grates and forced draft under the grate.

Other tests,† upon a return tubular boiler equipped with Coxe stoker, indicated the relative efficiencies of various fuels as presented in Table No. 42.

Influence of the Frequency of Firing.—While the rate at which any given coal should be fed to the furnace is largely

---

* "Furnace for Burning Small Anthracite Coals." Eckley B. Coxe. Trans. Am. Inst. Mining Engineers, vol. XXII.

† "Some Thoughts on the Economical Production of Steam, with Special Reference to the Use of Cheap Fuel, by a Miner of Coal." Eckley B. Coxe. Trans. New England Cotton Manufacturers' Association, April, 1895.

TABLE No. 41.

RESULTS OF TESTS OF PEA AND BUCKWHEAT COALS.

Items.	1. Oneida Pea Coal.	2. Oneida No. 1 Buckwheat.	3. Oneida No. 2 Buckwheat.	4. Oneida No. 3 Buckwheat.	5. Eckley No. 3 Buckwheat.
Pounds of water evaporated per pound of dry coal, actual conditions.................................	7.14	6.62	7.17	7.21	7.36
Pounds of water evaporated per pound of dry coal from and at 212°..................................	8.56	7.94	8.60	8.65	8.74
Pounds of water evaporated per pound of combustible from and at 212°...............................	10.14	10.06	10.57	11.12	11.10
Pounds of water evaporated from and at 212° per square foot of heating surface.....................	3.70	3.21	3.13	3.13	3.06
Pounds of coal per square foot of grate per hour...	13.63	13.58	11.40	11.34	9.44
Average temperature of escaping gases............	549°	543°	498°	503°	372°
Moisture in steam, per cent........................	2.2	2.0	1.9	1.9	....
Moisture in coal as fired, per cent.................	2.63	4.06	8.62	6.53	4.93
Per cent of ash...................................	15.60	20.10	18.71	22.27	21.3
Per cent of carbon in ash.........................	15.85	12.35	9.33	31.90	29.63
Average pressure of blast in inches of water in entrance chamber...............................	0.375	0.5	0.625	1.04	1.125
Analysis of coal. { Water at 225°..................	2.15	2.00	2.10	2.05	2.50
Volatile combustible matter,	5.10	4.90	5.45	5.42	5.00
Ash..................................	12.55	17.35	15.50	12.90	13.97
Carbon, fixed	80.20	75.75	76.95	79.63	78.53
Specific gravity	1.620	1.664	1.655	1.642	1.665
Sizing test. { Chestnut, over ⅞ inch round mesh.....	8.44	0.98	....	....	....
Pea coal, between ⅞ inch and 9/16 inch round mesh...........................	60.65	6.85	0.31	1.50	1.21
No. 1 Buckwheat, between 9/16 inch and ⅜ inch round mesh.............	21.70	57.72	4.76	4.58	2.60
No. 2 Buckwheat, between ⅜ inch and 3/16 inch round mesh.............	3.68	28.74	66.57	17.75	31.94
No. 3 Buckwheat, between 3/16 inch and 3/32 inch round mesh...........	1.40	2.39	19.87	45.95	49.57
Between 3/32 inch and 1/16 inch, sometimes allowed in No. 3 Buckwheat..................................	4.13	1.49	2.39	19.79	6.31
Dust through 1/16 inch round mesh ...	....	1.83	6.10	10.43	8.37
Slate test. { Pure coal, specific gravity below 1.70.	92.00	76.18	78.28	86.98	83.85
Slate and bone, specific gravity above 1.70..................................	8.00	23.82	21.72	13.02	16.15

TABLE No. 42.

RELATIVE EFFICIENCIES OF SMALL ANTHRACITE COALS.

Kind of Coal.	Pounds of Water per Pound of Coal from and at 212°.	Pounds of Water per Pound of Combustible from and at 212°.
Buckwheat.................................	8.77	11.07
Rice (No. 2 buckwheat)...................	9.05	11.18
Culm (pea, buckwheat, rice, barley, dust).	8.74	11.19
Barley (No. 3 buckwheat).................	8.39	10.89

EFFICIENCY OF FUELS. 107

dependent upon the character of the coal, nevertheless it is doubtless true that in most cases it is fed in too large quantities and at too long intervals. The natural result is a series of decided and almost critical changes in the condition of the fire, to be compared to the effect of eating a large amount of food once a day instead of smaller amounts at more frequent intervals. It is evident that efficiency, as regards the frequency of firing, is entirely dependent upon the fireman, and hence for favorable conditions the advantages of a mechanical stoker.

For the purpose of ascertaining, so far as possible, the relative results of different rates of firing, Burnat[*] conducted a series of experiments extending over eight weeks, with the same fireman and the same boiler.

The general results are presented in Table No. 43. The advantage of the smallest charge of 13 pounds over the maximum of 55 pounds is in the first series 3.03 per cent, and in the second 8.19 per cent. These results were obtained, notwithstanding the fact that with the more frequent firing the doors were more frequently opened. A boiler arranged so

TABLE NO. 43.

INFLUENCE OF FREQUENCY OF THE CHARGES OF COAL.

Kind of Coal.	Cubic Feet of Air at 62° per Pound of Coal.	Temperature of Feed-water.		Temperature of Hot Gases.		Pounds of Coal Consumed.			Per cent of Residue.	Pounds of Water Evaporated from and at 212° per Pound of Coal.
		Entering Feed-heater.	Entering Boiler.	Leaving Boiler.	Leaving Feed-heater.	Per Hour.	Per Square Foot of Grate.	Per Charge.		
Ronchamp, nut........	202	90°	213°	849°	421°	225	10.8	13.3	12.9	9.87
	202	87	224	840	426	225	10.8	26.6	13.4	9.59
	197	87.5	227	844	414	225	10.8	39.2	12.8	9.59
	202	86	226	835	421	225	10.8	55.4	12.8	9.58
Ronchamp, large and small......	226	87	230	779	396	225	10.8	55.0	16.1	8.91
	212	87	226	784	410	225	10.8	41.1	14.6	9.18
	201	87.5	229	795	410	225	10.8	28.0	14.7	9.38
	202	86	228	853	489	225	10.8	15.0	12.6	9.64

[*] Bulletin de la Société Industrielle de Mulhouse, vol. XLVI.

that the damper became closed or nearly so when the door was opened showed upon test an increased evaporation of 14 to 15 per cent due to this arrangement.

Loss on Account of Moisture in Coal.—Moisture in coal is an exceedingly variable quantity, depending upon the character of the coal, its temperature, and its previous exposure to the atmosphere. Under ordinary conditions its percentage varies from 1 to 5 per cent. Whatever its amount, it must all be raised to 212°, evaporated into steam, and the steam raised to the temperature of the escaping gases. It therefore has an important influence upon the theoretical heat-value of a given coal. Thus if one coal was composed of 80 per cent carbon, 15 per cent ash, and 5 per cent water, and another consisted of the same proportion of carbon, with 5 per cent ash and 15 per cent water, the theoretical calorific value—viz., 11,720 B. T. U.—would be the same, being directly dependent upon the amount of carbon. But in the first case the available heat (neglecting losses not due to water) would be 10,600 B. T. U., while in the second it would be 10,488 B. T. U., if the waste gases were assumed to escape at 500°. This points most clearly to the necessity of keeping coal as dry as possible if its maximum heating power is to be realized.

Loss on Account of Smoke.—The loss resulting from the formation of smoke is absolute; for it is equivalent to directly robbing the fire of a part of the fuel from which not only has no heating effect been secured, but upon which heat has actually been wasted in raising it to the temperature of the escaping flue-gases. Notwithstanding the prevailing impression as to the great losses due to the formation of smoke, the actual waste is comparatively insignificant, as is shown by the following results of carefully conducted experiments by Hoadley.[*]

During an entire week gas was drawn from the flue of the boiler under test, passed through a gas-meter and thence

---

[*] "Warm-blast Steam-boiler Furnace." J. C. Hoadley.

through a muslin strainer at the bottom of a vessel of water. When a sufficient quantity of the gases had been passed the water was evaporated, and the residuum was dried and weighed. The coal used was bituminous, of the following average composition:

Carbon.................	81.03	per cent
Hydrogen	3.84	"
Ash	7.19	"
Water...	0.63	"
Oxygen.................	4.49	"
Nitrogen...............	2.00	"
Sulphur................	0.82	"

The total quantity of coal burned during the week was 12,890 pounds, the total quantity of flue-gases reduced to 72° being 4,263,119 cubic feet, and the total amount of solid matter 42.63 pounds, as shown by the test. There was, therefore, present in solid form in the flue-gases only—

$$\frac{42.63}{12890} = .0033 = 0.33 \text{ per cent}$$

of the matter originally present in the coal. As the gray color of the substance thus recovered indicated that it was not more than half carbon, it is evident that under the conditions of the test the proportion of carbon which was actually carried off in black smoke was about one sixth of one per cent of the original coal.

Scheurer-Kestner and Meunier[*] passed flue-gases through asbestos, upon which the particles of carbon were deposited, and found that, with good fire and draft and an air-supply of 257 cubic feet per pound of coal, one half of one per cent of the carbon of the coal was lost as smoke-particles. In a second trial, with poor fire and draft and only 118 cubic feet of air per pound of coal, the deposited carbon was one per cent of the total contained in the coal. This latter is to be

---

[*] Bulletin de la Société Industrielle de Mulhouse, 1868, 1869.

taken as a maximum, the conditions being decidedly adverse, and is more than would be produced in ordinary practice. The inference from these experiments must be that the average loss of carbon in the solid form as smoke may be taken not to exceed one half to three quarters of one per cent.

Loss on Account of Carbonic Oxide.—The loss of efficiency which ensues from the escape of carbonic oxide unconverted into carbonic acid is due to the much smaller amount of heat given out upon the incomplete combustion of carbon into carbonic oxide. While carbon burned to carbonic acid generates 14,650 B. T. U., the same quantity burned to carbonic oxide gives out only 4400 B. T. U. For every pound of carbon which passes off in the form of carbonic oxide there is, therefore, a loss of $14,650 - 4400 = 10,150$ B. T. U., or

$$\frac{10150}{14650} = 0.6928 = 69.28 \text{ per cent.}$$

The ultimate effect of the formation of carbonic oxide upon the total heat of combustion can be best illustrated by reference to the calculation under "Heat of Combustion" in Chapter III. If, instead of the entire 0.80 pound of carbon having been perfectly burned, only 0.70 pound had entered into combustion with oxygen to form carbonic acid, and the remaining 0.10 pound had entered into combustion as carbonic oxide, the calculation would have been as follows:

Heat, in B. T. U.,

$$= 14650 \times 0.70 + 4400 \times 0.10 + 62100\left(0.05 - \frac{0.027}{0.8}\right)$$

$$= 13590 \text{ B. T. U.};$$

showing that there would be a loss in the calorific value of the coal of about 7 per cent. Had only half of the carbon been completely burned, the loss would have been about 18 per cent. But such a loss even as that first instanced does not usually occur continuously in any well designed and operated boiler-furnace. In fact, the loss from this source appears to be largely overestimated in most cases.

In Hoadley's warm-blast furnace-tests the carbonic oxide "never in the daytime exceeded half of one per cent, and rarely exceeded half that small quantity when the dampers were open, for six weeks together." The proportions revealed by Bunte's tests have already been shown in Table No. 15.

From the analyses of the chimney-gases of 124 boiler-tests made at the Industrial Exhibition at Düsseldorf in 1880,* the average amount of carbonic oxide by volume was found to be 0.747 per cent. In all cases but one bituminous coal was used. The air-supply was almost constant; the minimum amount of nitrogen shown in any of the tests being 79.3 per cent, and the maximum 84.08 per cent, but ranging in most cases between 80 and 81 per cent. Omitting four analyses in which the carbonic oxide ranged from 3 to 5.3 per cent, the average presence of this gas in the remaining 120 tests appears to have been 0.637 per cent.

There are some conditions of boiler-practice, however, in which such good results do not obtain, and in which more or less serious losses may occur, largely due to an insufficiency of air. It should be clearly understood, however, that the same amount of air is not always required. When the coal upon the grate is thoroughly ignited the minimum supply is necessary, but when the fire is suddenly thickened and cooled by additional coal there is a demand for additional supply for the purposes of combustion, together with a tendency to clog the passages through which the air has previously passed, and thereby to prevent complete combustion at the surface of the fire. At this time, for perfect conditions, more air should be admitted under the influence of more intense draft.

The effect of excessive firing is practically equivalent to the reduction of draft and air-supply for the regular amount

---

* "Die Untersuchungen an Dampfmaschinen und Dampfkesseln, und an einigen Rheinischen und Westfälischen Kohlensorten auf der Gewerbe-Ausstellung in Düsseldorf in 1880," herausgegeben von H. v. Reiche und F. Böcking, Aachen, 1881. Verlag von J. A. Meyer.

of coal, and is conducive to the production of carbonic oxide. This was well exemplified by Hoadley in a special test,* the results of which are presented in Table No. 44. As is evident

TABLE No. 44.

CARBONIC OXIDE PRODUCED BY EXCESSIVE FIRING.

Time.	Pounds of Coal Thrown on Grate.	Carbonic Acid in Chimney Gases.	Carbonic Oxide in Chimney Gases.	Ratio of Carbon in Carbonic Oxide to Total Carbon.	Pounds of Air per Pound of Coal.	Pounds of Coal Burned Each Half-Hour.	Ratio of Loss by Carbonic Oxide to Full Power of Coal.
h. m.	Pounds.	Per Cent CO_2.	Per Cent CO.	Per Cent.	Pounds.	Pounds.	Per Cent.
6:15 a.m.	200						
6:45	200						
7:15	200						
7:45	200						
8:15	200						
8:45	200						
9:00		5.12	2.54	43.80	33.2	83.81	27.84
9:15	200						
9:30		5.55	2.99	45.85	29.5	93.75	29.14
9:45	200						
10:00		7.79	3.99	44.63	21.4	129.24	28.37
10:15	200						
10:30		7.70	4.61	48.47	20.1	137.60	30.81
10:45	200						
11:00		7.82	4.70	48.57	19.8	139.68	30.88
11:15	200						
11:30		8.01	4.81	48.55	19.3	143.30	30.86
12:00 m.					19.3	143.30	
12:30 p.m.		15.21	0.25	2.52	19.3	143.30	1.60
12:45	200						
1:00					20.05	137.94	
1:30		14.11	0.21	2.28	20.8	132.96	1.49
2:00					21.05	137.94	
2:30		13.62	0.33	3.67	21.3	129.85	2.31
2:45	200						
3:00		14.50	0.48	4.95	19.4	142.56	3.14
3:30		13.18	0.29	3.34	22.0	125.34	2.12
3:45	200						
4:00		14.96	0.38	3.84	19.3	143.30	2.44
4:30		14.18	0.41	4.35	20.3	136.25	2.76
4:45	200						
5:00		13.01	0.41	4.72	22.0	125.34	3.00

Mean quantity of air.......................... 21.653
Mean of all but first two.................... 20.36
Mean ratio of loss ; first six, per cent............................ 29.65
Mean ratio of loss ; last eight, per cent........................... 2.36

* "Warm-blast Steam-boiler Furnace." J. C. Hoadley.

from the items in the first and second columns, the firing was at first very rapid, decreasing as the day passed. During the morning 200 pounds were fired every half-hour, while during the afternoon the same amount was fired hourly. Gas-samples were not taken until 9 A.M., after which time the excess of carbonic oxide is noticeable until the slower firing began, when the carbonic oxide is immediately reduced, while the carbonic acid is practically doubled. The ratio of loss by carbonic oxide, indicated in the last column, is of special interest.

Admission of Air Above the Fire.—Since the days of C. Wye Williams and his famous work on the "Combustion of Coal and the Prevention of Smoke," the introduction of air above or beyond the fire has been one of the favorite methods adopted in the attempt to perfect the combustion of coal and prevent the appearance of smoke. In so far as the admission of air for the purpose of perfecting the combustion is concerned, the results in the way of water evaporation per pound of fuel, independently of the prevention of smoke, should naturally be looked to as the true indication of its efficiency. That, under certain conditions, increased efficiency can be thus obtained is evidenced in many tests. As a rule the gain is not large, but is greatest with bituminous coals, whose large percentage of volatile constituents tends more readily to the formation of both carbonic oxide and smoke.

Table No. 45, compiled from carefully conducted tests by Barrus,[*] serves to show the percentage of both gain and loss by such contrivances, the character of which is indicated under the designating letters, as follows:

A. Air conducted direct from outside the setting to the interior of the bridge-wall and discharged therefrom through perforations in its top covering. The air thus supplied mingles with the lower strata of burning gas as it skims over the bridge.

---

[*] "Boiler-tests." George H. Barrus.

TABLE No. 45.

EFFECT OF ADMITTING AIR ABOVE THE FIRE.

Designation	Kind of Coal.	Gain.		Loss.	
		Per Pound of Coal.	Per Pound of Combustible.	Per Pound of Coal.	Per Pound of Combustible.
A	Cumberland..	5.9	6.2	....	....
A	Anthracite, broken	....	....	0.0	1.0
A	Two parts pea and dust and one part Cumberland.	....	....	2.0	4.7
B	Cumberland.	8.4	8.0	....	....
B	Anthracite, broken.............................	1.9	3.7	....	....
C	Two parts pea and dust and one part culm.......	2.0	0.0	....	....
D	Two parts anthracite screenings and one part Cumberland..............................	....	....	4.3	2.3
E	Three parts pea and dust and one part Cumberland	....	....	4.5	4.7
F	Nova Scotia..	1.0	1.5	....	....

B. Air supplied first to a pipe laid in the bridge-wall and thence to perforated cast-iron globes resting upon its top. A more thorough mixture of air and gases is thus secured. A jet of steam is employed to increase the volume of air which would be drawn in by natural means, and the steam thus supplied mingles with the air.

C. Air supplied through perforations in the top of the bridge-wall and in the sides of the furnace after first passing through a series of passages running lengthwise of the walls of the setting.

D. Air admitted through perforated tiles in the sides of the furnace of a vertical tubular boiler, after having passed up and down through ducts in the walls.

E. Air admitted in similar manner in the case of a horizontal tubular boiler, after having passed through side-wall heating-flues, as in the case of D.

F. Air supplied through perforations in the top of the bridge-wall and inside of furnace.

Barrus concludes from these tests that "a considerable advantage attends the admission of air above the fuel when bituminous coal is employed, the amount of gain depending somewhat upon the method employed. There is no advantage in the system where mixtures of anthracite screenings and bituminous coal are used, if carried out according to either the first or fourth methods (A and C, as here designated), and, finally, little or no benefit is derived when anthracite coal is burned."

It will be observed that four of the boilers here designated were provided with certain devices for heating the air before its admission to the fire. The gain in efficiency, by such an arrangement, is at the best but slight, and when the attempt is made to heat by such means all of the air supplied both below and above the fire, this plan is found to be totally inadequate. Such increase in temperature of the air-supply as may be secured by proper and adequate appliances properly concerns the efficiency of the boiler, under which heading it will be discussed. Although it is undoubtedly true that heating the air intensifies the chemical affinities between the air and the fuel, it is doubtful if under ordinary conditions the effect is sufficient to be noticeable in boiler-practice. This is, however, independent of the economy resulting from the abstraction of waste heat from the flue-gases.

The best results from the admission of air above the fire appear to be secured when it is discharged in fine jets into the furnace-chamber. For the accomplishment of such results other draft than that of the chimney is necessary. It is thus that mechanical draft becomes an important factor, both in thus furnishing the required air-supply and in overcoming the added resistance which results from any attempt to preheat the air.

Loss on Account of Excess of Air.—The method of calculating the theoretical amount of air required for the complete combustion of any fuel has already been explained. If less than this amount is supplied, a certain portion of the carbon passes off unconsumed and forms smoke; while a part of the remainder, being insufficiently supplied with oxygen, forms carbonic oxide, the product of incomplete combustion. If the air be supplied in excess of that necessary for perfect combustion, there is a definite loss, which is twofold in its character: First, the excess of air entering the furnace is heated by the burning fuel, thereby lowering the temperature of the mixture of gases and air below that which would prevail if the gases only were present. As a consequence, the rate of absorption of heat by the water is reduced, for it is dependent upon the difference in temperature between the water and the gases. Second, owing to larger volume and higher velocity the temperature of the mixture of gases and air escaping to the chimney is higher than would be the case if there were no excess of air; while the increased volume is such that the total amount of heat thus carried away, without exerting any useful effect, is greatly increased. In other words, paradoxical as it may seem, the larger the volume of air supplied, the higher will be the temperature of the escaping gases.

As has already been stated, it is a practical impossibility so to distribute the amount of air theoretically required for the perfect combustion of a given fuel as to secure the ideal result. In practice, notwithstanding the fact that all excess means loss of efficiency of the fuel, it is necessary to supply enough additional air to insure complete combustion even if a loss is occasioned thereby. What this excess should be, from an economical standpoint, can only be determined by practical experiment; and it will be found to vary with the character of the fuel, its rate of combustion, the temperature of the supply, and the intensity of the draft.

For the purpose of illustration an anthracite coal may be taken, containing 80 per cent of carbon, with an amount of inherent oxygen and hydrogen so small in quantity and of so little effect that, under the circumstances, it may be neglected. The amount of air required for the complete combustion of one pound of carbon having already been shown to be very nearly 11.51 pounds, that necessary for the combustion of one pound of this coal is, therefore, $11.51 \times 0.8 = 9.208$ pounds, as is also evident by the following calculation, which differs slightly because of greater refinement:

Carbon................	0.8	Oxygen...............	2.133
Oxygen................	2.133	Nitrogen..............	7.071
Carbonic acid	2.933	Air...................	9.204
Nitrogen..............	7.071	Carbon...............	0.80
Products	10.004	Products.............	10.004

The curves presented in Fig. 7 show clearly the effect upon the weight and ideal temperature of the gaseous products of combustion and their relative volume, resulting from the supply of air in various amounts in excess of that theoretically necessary for combustion.

Evidently, as the total volume of gases is heated in each case by the products of only one pound of coal, the total heat is constant; but because of the greater volume absorbing that heat the ideal temperature, as shown by curve A, decreases as the air-supply increases.

The relative weight of the products of combustion, as indicated by curve B, obviously is proportional to the initial weight of the air supplied plus a constant represented by the portion of the fuel, 0.8 pound in this case, which has united with the oxygen to form carbonic acid.

The decrease in the ideal temperature which follows from an increased admission of air results in less expansion and consequently greater density of the products of combustion. For this reason the relative weight, as indicated by curve C, remains practically constant.

For simplicity, the effect of the greater specific gravity of carbonic acid has been neglected in the calculations, and the products of combustion have been taken as of constant density, at constant temperature, with varying amounts of air supplied.

It is generally accepted that in the case of a steam-boiler

the amount of heat transmitted from the gases to the water, per degree difference between the gases and the boiler plates and tubes, is practically uniform for various differences of temperature. In other words, it is practically proportional to the difference in temperature. It is, further, evidently true

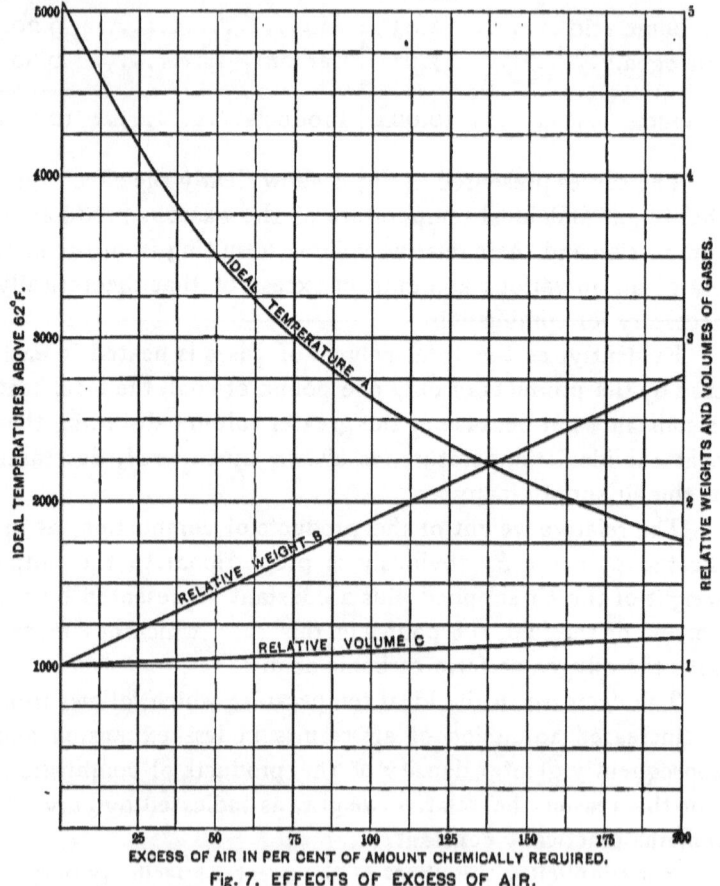

Fig. 7. EFFECTS OF EXCESS OF AIR.

that in the case of moving gases the amount of heat transmitted will be approximately proportional to the time they remain in contact with the given surface. The dual influence thus exerted upon the final temperature of the gases as they leave the boiler is somewhat complicated in its character.

As the products of combustion pass across the heating-surface and are cooled, their volume, and hence their velocity, decreases, and as their temperature approaches that of the admitted air, so does their volume. In other words, the average rate of flow becomes more nearly proportional to the original air-volume. As a consequence, the cooler gases resulting from the admission of a larger volume of air have less time in which to give up their heat, and are, therefore, cooled less in proportion to their temperature than are the hotter gases resulting from the admission of a smaller air-volume. In addition, these cooler gases, because of the less difference between their temperature and that of the boiler, give up their heat less rapidly; the final result of these two influences, by which the transmission of heat is retarded, being that the gases accompanying the larger admission of air leave the boiler at a higher temperature than those which result from the admission of a smaller amount. Evidently, under these conditions the evaporative power of the boiler per pound of fuel must be decreased, but with the increased air-supply a greater total amount of coal may be burned.

Of course it is to be understood that these conditions exist and results ensue because of the relative proportions of heating-surface and grate-surface in the ordinary boiler. It has just been rendered clearly evident by Fig. 7, that when the heat-abstracting influence of the boiler is eliminated, the temperature of the escaping gases, which is in effect the temperature of the furnace, decreases as the excess of air increases. But the intermediary boiler serving, as it ordinarily does, to to reduce the temperature of the gases to 400° or 500°, cools them sufficiently to permit of securing the unexpected result of an increasing flue-gas temperature corresponding to the augmented volume of air supplied to the fire. It is, therefore, manifest that the lower the normal temperature of the escaping gases, the greater will be their increment in temperature resulting from the admission of more air.

Burnat conducted a series of experiments, showing that

the temperature of the escaping gases increased with the supply of air. The boiler, provided with special heaters, had a heating-surface of 475 square feet, with a grate of 18 square feet; the quantity of coal consumed was 293 pounds per hour,

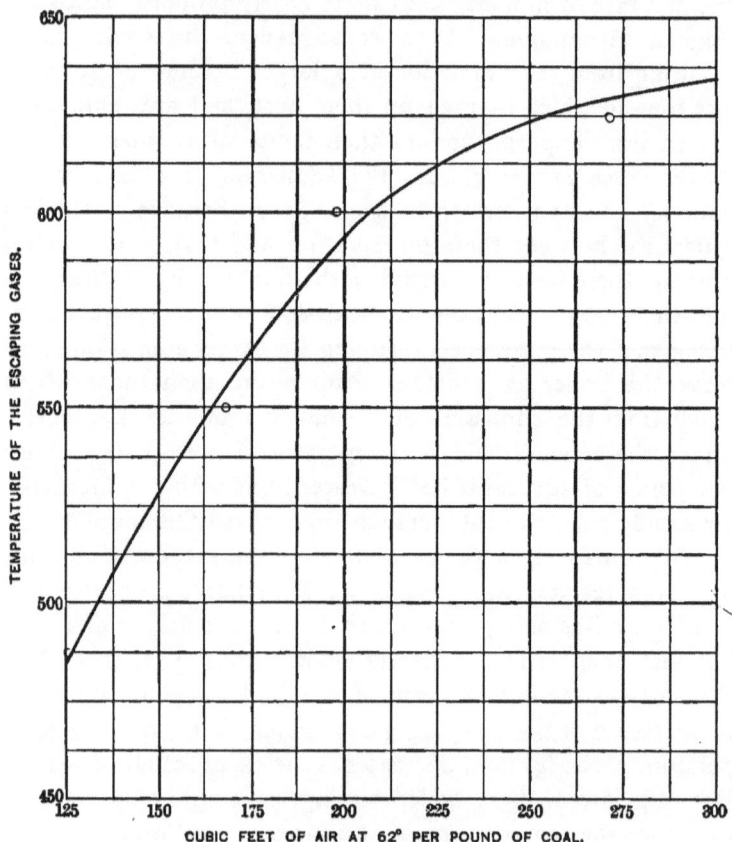

Fig. 8. EFFECT OF INCREASED AIR SUPPLY ON TEMPERATURE OF THE ESCAPING GASES.

or 16 pounds per square foot of grate. These results are given in Table No. 46 and also presented in the curve Fig. 8, which serves to show their general trend. If the temperatures had been determined and plotted for larger air-volumes, it is evident that the curve would have gradually approached the base-line, for with an infinite volume of air the ideal temperature must coincide with the initial temperature.

TABLE No. 46.

EFFECT OF INCREASED AIR-SUPPLY UPON TEMPERATURE OF ESCAPING GASES.

Day.	Cubic Feet of Air at 62° per Pound of Coal.	Average Temperature of the Gases leaving the Boiler.
First	272	624°
Second	198	601
Third	168	550
Fourth	124	487

The theoretical loss of efficiency, when air is supplied in excess and the products of combustion escape at different temperatures above the atmosphere, is exemplified in Table No. 47, the coal having a heat-value of 11,720 B. T. U. This relates to the combustion of one pound of coal. An air-

TABLE No. 47.

LOSS OF EFFICIENCY DUE TO EXCESS OF AIR, AND TEMPERATURE OF ESCAPING GASES ABOVE ATMOSPHERE.

Excess of Air in per cent of that Chemically Required.	Temperature of Escaping Gases above Atmosphere.												
	300°.		350°.		400°.		450°.		500°.		550°.		
	Total B. T. U. in Gases.	Loss in per cent of Total Heat Value of Coal.	Total B. T. U. in Gases.	Loss in per cent of Total Heat Value of Coal.	Total B. T. U. in Gases.	Loss in per cent of Total Heat Value of Coal.	Total B. T. U. in Gases.	Loss in per cent of Total Heat Value of Coal.	Total B. T. U. in Gases.	Loss in per cent of Total Heat Value of Coal.	Total B. T. U. in Gases.	Loss in per cent of Total Heat Value of Coal.
0	695	5.9	812	6.9	928	7.9	1044	8.9	1160	9.9	1276	10.9
50	1016	8.7	1185	10.1	1354	11.5	1524	13.0	1693	14.4	1862	15.9
75	1176	10.0	1372	11.7	1568	13.3	1764	15.0	1959	16.7	2155	18.4
100	1336	11.4	1558	13.3	1781	15.2	2003	17.0	2226	19.0	2448	20.9
125	1495	12.7	1745	14.9	1994	17.0	2243	19.1	2492	21.2	2742	23.4
150	1655	14.1	1931	16.5	2207	18.8	2483	21.1	2759	23.5	3035	25.9
175	1815	15.5	2118	18.0	2420	20.6	2715	23.1	3025	25.8	3328	28.4
200	1975	16.8	2304	19.6	2633	22.4	2957	25.2	3291	28.0	3621	30.8

supply 100 per cent in excess of that chemically required, and a temperature of escaping gases 450° above the atmosphere, represents fairly well the ordinary conditions of boiler-practice with chimney-draft, under which it will be noted that the loss is no less than 17.0 per cent. A chimney requires a high and wasteful temperature of gases to produce the draft, but this is unnecessary with mechanical means for draft-production. A moderate reduction to 75 per cent excess air-supply and 300° would show an economic gain of 7 per cent.

As bearing upon the evaporation of the boilers, the results embodied in Table No. 48 emphatically indicate the loss due to an increase in the air-supply. These tests, by Burnat,* were upon boilers of various types. The air chemically required for perfect combustion was 130 cubic feet per pound in the case of the Ronchamp coal.

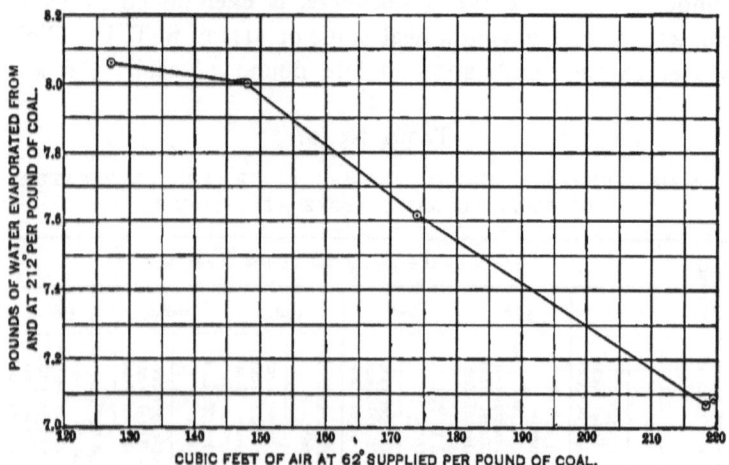

Fig. 9. FULL EFFICIENCY WITH CONSTANT COAL CONSUMPTION AND VARIABLE AIR SUPPLY.

A comparison of the efficiencies with different amounts of air but with constant coal consumption is indicated by the curve Fig. 9. This is plotted for the Ronchamp mixed coal,

---

* Bulletin de la Société Industrielle de Mulhouse, vol. xxx.

TABLE No. 48.

EFFECT OF AIR-SUPPLY ON THE EFFICIENCY OF FUEL.

Boiler.	Coal.	Relative Efficiency.	Coal per Hour.	Air at 62° supplied per Pound of Coal.	Water Evaporated from and at 212° per Pound of Coal.
		Per cent.	Pounds.	Cu. Ft.	lbs.
Heating-surface = 513 square feet. Grate-surface = 14.2 square feet.	Ronchamp mixed.	100 100 107 112 113	330 330 330 330 330	219 216 174 148 127	7.09 7.08 7.62 8.00 8.06
Heating-surface = 301 square feet.	Sarrebrück slack, very inferior.	100 104 108 112 112 108 110	284 285 276 257 242 237 234	222 229 200 145 153 207 126	5.46 5.67 5.93 6.11 6.13 5.92 6.01
Heating-surface = 475 square feet. Grate-surface = 18 square feet.	Sarrebrück slack, very inferior.	100 108 112 114 111 101	367 370 375 361 367 316	290 264 190 141 196 121	5.26 5.67 5.88 6.03 5.86 5.32
Heating-surface = 291 square feet.	Half Ronchamp slack and Sarrebrück slack.	100 108 107 110	280 263 259 260	190 169 152 123	6.60 7.10 7.09 7.26

the only series of tests in which there was no variation in the amount of coal consumed per hour. Although thus restricted to displaying only one set of results, it very clearly shows the decrease in efficiency which follows an increase in the air-supply. The effect of the excess of air in not only increasing the temperature of the escaping gases, but also their quantity, readily accounts for this reduction in evaporative efficiency.

From the results given in the table it would appear that the best practice consisted in keeping the volume of air admitted as small as possible. But the result of such supply, although increasing the efficiency, even though imperfectly burning the coal, would cause a dull fire with abundant smoke and meet with serious practical objection on account of the difficulty of maintaining it under varying conditions of demand. It would, in fact, be a fire requiring the utmost care and attention on the part of the fireman, and liable to fail to raise steam when suddenly required.

In all boiler practice there is no factor more subtle in its influence than that which has just been discussed. Because air is invisible, its weight or volume supplied to the fire not easily determined, and even the intensity of the draft but seldom indicated, it is natural that proper regulation of the supply should not be readily secured. Except in so far as evidenced by the formation of smoke, a lack of air and draft is not ordinarily appreciated by the fireman, and great waste may continue to occur on account of incomplete combustion and the discharge of its imperfect result, carbonic oxide. On the other hand, as elsewhere shown, the arrangements may be such as to render the small volume of air one of the most important elements in securing increased efficiency in the evaporation per pound of coal.

Summary of Influences Affecting the Efficiency of Fuel. —The relative importance of the principal influences which have just been discussed and the manner in which they are exerted may be very clearly shown by an analysis of the results of combustion of a given quantity of coal, in the manner adopted by Coxe.* For the purpose of illustration there has been taken 100 pounds of anthracite coal of the composition shown in the accompanying Tabular View. It is

---

* "Some Thoughts on the Economical Production of Steam, with Special Reference to the Use of Cheap Fuels, by a Miner of Coal." Eckley B. Coxe. Trans. New England Cotton Manufacturers' Association, April, 1895.

EFFICIENCY OF FUELS. 125

Tabular View

SHOWING RESULTS OF COMBUSTION OF 100 POUNDS OF ANTHRACITE COAL WITH TWICE THE THEORETICAL AMOUNT OF AIR REQUIRED.

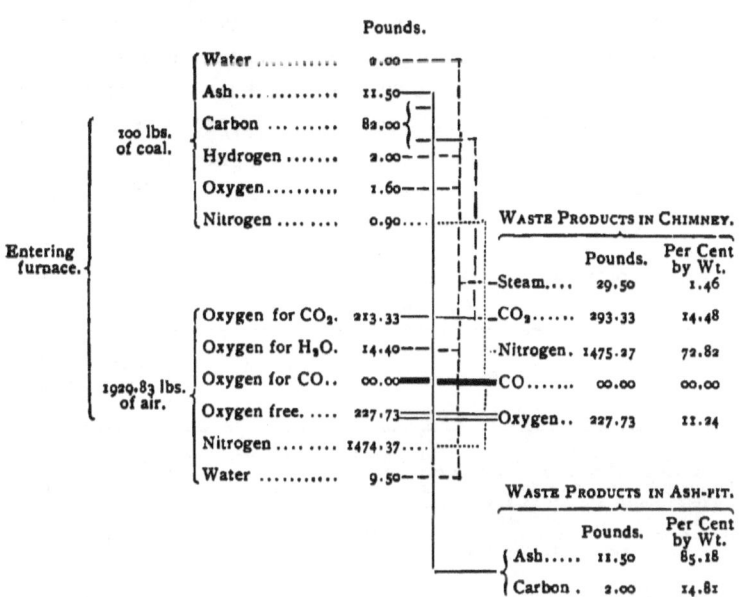

Total Heat of Fuel.

Weight of $C \times 14{,}650 = 82 \times 14{,}650 = \ldots\ldots\ldots\ldots\ldots$ 1,201,300 B. T. U.

Weight of $H - \left(\frac{\text{Wt. of O}}{8}\right) \times 62{,}100 = 2 - \left(\frac{1.6}{8}\right) \times 62{,}100 = 111{,}780$ "

 1,313,080 "

Heat Generated.

$80 \times 14{,}650 = \ldots\ldots\ldots\ldots\ldots\ldots\ldots\ldots\ldots\ldots\ldots\ldots$ 1,172,000 B. T. U.

$2 - \left(\frac{1.6}{8}\right) \times 62{,}100 = \ldots\ldots\ldots\ldots\ldots\ldots\ldots\ldots$ 111,780 "

 1,283,780 " assumed that double the theoretical amount of air is supplied, that the atmosphere contains a normal amount of moisture, that two pounds of carbon remain unconsumed and pass to the ash-pit, that neither carbonic oxide nor smoke is formed, that the coal and air have an original temperature of 60° when they enter the furnace, that the chimney-gases have a temperature of 500°, and that the ash has a temperature of 450°. The total heat of 100 pounds of the fuel is shown to be 1,313,080 B. T. U., but, owing to direct loss to the ash-pit, the heat generated amounts to only 1,283,780 B. T. U. The Tabular View indicates the relative weights of the various products of combustion, and the accompanying Table No. 49

TABLE NO. 49.

HEAT LOSSES INCIDENT TO THE COMBUSTION OF 100 POUNDS ANTHRACITE COAL.

Heat-losses.	Number of B. T. U.	Per cent of Total Heat of Fuel.
By water = [(212 − 60) × wt.] + 965.7×wt. + [sp. heat × (500−212) × wt.]	37012.5	2.83
By carbonic acid = wt. × sp. heat × (500 − 60)................................	27994.2	2.13
By nitrogen = wt. × sp. hea × (500 − 60)........................	158452.8	12.07
By free oxygen = wt. × sp. heat × (500 − 60)...............	21973.6	1.67
By ash = wt. × sp. heat × (450 − 60).................................	1105.7	0.08
By carbon in ash = wt. × sp. heat × (450 − 60) + wt. × 14650...........	29488.3	2.24
By carbonic oxide = wt. × sp. heat × (500 − 60) + wt. × 4400............	....	....
Total heat lost exclusive of loss by radiation	276027.1	21.02
Theoretically possible evaporation in pounds of water from and at 212° per pound of combustible utilized... ....;	12.73	
Theoretically possible evaporation in pounds of water from and at 212° per pound of fuel utilized........	·10.44	

shows the heat-losses from these sources, together with their methods of calculation. It is evident that under the assumed conditions 21.02 per cent is irretrievably lost, and that, neglecting the loss by radiation from brickwork, the efficiency of the fuel is 78.98 per cent. With the given fuel these losses can only be lessened by decreasing the air-supply, preventing the loss of some of the carbon to the ash-pit, and by proper means lowering the final temperature of the escaping gases. The total loss through moisture in coal and air, ash in coal, and carbon in ash amounts to 5.15 per cent.

Commercial Efficiency of Coals.—The cost of producing a given amount of steam is the ultimate criterion by which the efficiency of any fuel must be judged. The commercial efficiency not only concerns the amount of water evaporated by a pound of fuel, but is directly dependent upon the following items of expense: *first*, interest, rent, taxes, and insurance on the cost, and the depreciation of the plant; *second*, repairs; *third*, cost of water from which the steam is generated; *fourth*, labor; *fifth*, getting rid of the ashes; *sixth*, cost of the fuel in the boiler-house.

The cost of water may be considered practically constant in the comparison of fuels in a given boiler-plant, but the items of interest, depreciation, and repairs are to a great extent directly dependent upon the fuel; for with the cheaper fuels more expensive contrivances are usually necessary, and under the same conditions the output of the plant decreases with the quality of the fuel. This latter fact is quite clearly shown in Table No. 41, in which the water evaporated per square foot of heating-surface grows less as the quality of the fuel degenerates. It is still further indicated by Table No. 50, in which the horse-power of various boilers, tested by Barrus[*] under

TABLE No. 50.

EFFECT OF QUALITY OF FUEL UPON OUTPUT OF BOILER.

Kind of Fuel.	Rated Horse-power of Boiler.					
	54	74	87	129	140	270
Cumberland	60.0	143.8	105.4		94.0	214.6
Anthracite, broken	53.9	105.5	84.0	192.3	103.8	196.1
Pea				149.2		
One part screenings and one part Cumberland		95.1				
Three parts screenings and one part Cumberland						204.8
Two parts pea and dust and one part Cumberland	38.5		82.2		118.1	
Forty-four parts pea and dust and thirty-seven parts culm				157.1		

[*] "Boiler-tests." George H. Barrus.

somewhat different draft conditions, is compared with their rated power when using various kinds of coal. This decreased power is largely due to the reduced combustion per square foot of grate, which may be counteracted to a certain extent by increasing the grate-surface.

The largest and most important factor, however, is the cost of the fuel itself, which should be measured, not by the number of pounds, but by the available heat-units obtained for a given price. In this cost are properly included the transportation charges, the expense of getting the coal into the boiler-house, and putting it into the furnace, as well as taking out and carrying away the ash. Practically all of these costs are directly dependent upon the weight of the coal, regardless of the number of heat-units it is capable of developing. But the net cost depends, not upon the number of units in the coal, but upon the number that can be utilized under the given conditions. The haulage of ashes becomes so important in some cases that it is found more economical to pay a higher price for a coal containing less ash rather than go to the necessary expense of teaming to a considerable distance the ashes from a poorer fuel.

As the cost of transportation of the coal is practically dependent upon the weight and independent of the character of the coal, the proportional difference in price which may rule at the mine may be almost extinguished when an equal charge is added for the transportation of each. Thus No. 2 buckwheat may prove a very economical fuel when utilized at the mine, where it may be purchased at 25 cents per ton, although its direct calorific efficiency is low and it requires a special form of furnace for satisfactory combustion. No. 1 buckwheat coal costing 50 to 60 cents under the same conditions may, however, prove to be a close rival because of its higher efficiency and the greater ease with which it may be consumed. But this difference of 100 to 120 per cent in cost becomes reduced to a difference of only 11 to 15 per cent when a transportation charge of $2.00 per ton is added to each. Both of these coals, although requiring a larger boiler-plant for the same aggregate evaporative results, may, after transportation charges are added, prove on the whole more economical than larger-sized anthracite of higher efficiency but at higher prices.

A comparison of the losses shown in Table No. 49, reduced to the commercial relation when different kinds of coal are used serves to make the preceding clear. This comparison is presented in Table No. 51, showing the relation between

TABLE No. 51.

COMMERCIAL VALUE OF THE LOSSES INCIDENT TO BURNING 100 POUNDS OF COAL.

Losses.	Kind and Price of Coal.			
	Buckwheat Coal.		Pea Coal.	
	$0.50	$2.50	$1.50	$3.50
By water.	$0.00063	$0.00315	$0.00189	$0.00442
By carbonic acid	0.00048	0.00238	0.00143	0.00333
By nitrogen	0.00269	0.01347	0.00808	0.01886
By free oxygen	0.00037	0.00186	0.00112	0.00261
By ash	0.00002	0.0009	0.00005	0.00012
By carbon in ash	0.00050	0.0025	0.00150	0.00350
Total loss, not including radiation from brickwork.	0.00469	0.02245	0.01407	0.03284
Actually utilized, including radiation from brickwork	0.01763	0.08816	0.05289	0.12341
Cost of fuel to evaporate 100 pounds of water from and at 212°.	0.00214	0.01069	0.00641	0.01497

an anthracite buckwheat costing 50 cents per ton at the mines, the same coal when transportation charges of $2.00 have been added, and pea coal costing $1.50 at the mines, with the same expense for transportation as the buckwheat. In each case the coal is assumed to be burned with double the theoretical amount of air, while the loss by radiation, which might amount to from 4 to 20 per cent, has not been considered. This loss, if known, should be deducted from the amount here given as actually utilized. The cost is based in the four cases upon the theoretical evaporation of 10.44 pounds of water from and at 212° per pound of coal, as given in Table No. 49. In practice these results would undoubtedly have to be corrected, relatively as well as directly, because of the higher evaporation probable with the pea coal.

Based upon the cost of the coal and firing, the following will serve to illustrate the economy that may be secured by burning a cheaper fuel of lower efficiency. In a certain plant equipped with horizontal tubular boilers and down-draft furnaces, an evaporation of 8.5 pounds of water from and at 212° was regularly obtained. The coal used was Illinois slack, costing $1.40, delivered under the boilers. This makes the cost about 8.3 cents per 1000 pounds evaporated under these conditions. Had the best picked anthracite coal, costing $4.80 per ton, under existing market rates, been used, and had it evaporated the generous amount of 12 pounds of water from and at 212°, the cost would have been 20 cents per 1000 pounds.

The relative evaporative values of different coals, as determined by Barrus[*] and presented in Table No. 52, will serve

TABLE No. 52.

RELATIVE EFFICIENCIES OF VARIOUS COALS.

Kind of Coal.	Water Evaporated from and at 212° by 1 Pound of Dry Coal.	Relative Efficiency in per cent. Cumberland = 100.	Cost of Coal per Ton.	Fuel Cost of Evaporating 1000 lbs. of Water from and at 212°.	Relative Efficiency in per cent Measured by Cost to Evaporate 1000 lbs. Cumberland = 100.
Cumberland.....................	11.04	100	$3.75	$0.1698	100
Anthracite, broken.............	9.79	89	4.50	0.2297	74
Anthracite, chestnut...........	9.40	85	5.00	0.2660	64
Two parts pea and dust and one part Cumberland.........	9.38	85	2.58	0.1375	123
Two parts pea and dust and one part culm................	9.01	82	2.58	0.1432	119
Anthracite, pea................	8.86	80	4.00	0.2259	75
Nova Scotia culm..............	8.42	76	2.00	0.1187	156

as an excellent basis for comparison upon a broader scale. The prices per ton were those ruling in Boston at the time of

---

[*] "Boiler-tests." George H. Barrus.

this writing. Measured solely by the fuel cost of evaporating 1000 pounds of water, the advantage is manifestly with the less efficient but cheaper coal, while the high-priced anthracites show the smallest return.

It is true that as the quality of the coal grows poorer and the size of the particles less, it becomes more necessary to provide greater intensity of draft, preferably by mechanical means, and some special form of grate or stoker for its proper burning. If the latter means are not provided, the labor account for firing will probably be increased. But even with due allowance for these items it is usually possible to show a commercial advantage in the use of low-grade fuels if burned under the proper conditions.

The possible annual savings in fuel-cost alone with coals of low grade is well evidenced in Table No. 53.

TABLE No. 53.

ANNUAL SAVINGS RESULTING FROM BURNING CHEAPER FUEL IN 1000 H.P. BOILER-PLANT.

Water Evap. from and at 212° per lb. of Coal.	Cost per Ton.													
	$0.50	0.75	1.00	1.25	1.50	1.75	2.00	2.25	2.50	2.75	3.00	3.25	3.50	3.75	4.00
11.00												4892	3669	2446	1223	0000
10.50											5474	4193	2912	1630	349
10.00										6115	4770	3424	2079	734
9.50							8240	6823	5407	3991	2575	1159
9.00						9105	7610	6115	4621	3126	1631	136
8.50					10074	8491	6909	5326	3743	2160	578
8.00				11180	9478	7797	6115	4433	2752	1070
7.50			12393	10599	8805	7012	5218	3424	1630
7.00	15724	13803	11881	9939	8037	6115	4193	2272	350

This table has been calculated for a 1000 H.P. power plant and is based upon 312 working days of 10 hours each. Cumberland coal costing, in round numbers, $4.00 per ton and evaporating 11 pounds of water from and at 212° has been taken as unity. At this rate the total annual cost for coal would be $19,568. If the assumption be made that, for instance, a coal costing $2.50 per ton and evaporating only

9 pounds of water is substituted, the annual saving is shown to be $4621. If a fan be introduced to produce the requisite draft, and if it required for its operation even 1½ per cent of the coal fed to the boilers (assuming that the exhaust-steam from the fan-engine is not utilized), this would amount to only $224 in value, which, if charged against the saving, would still leave it at $4397, an amount sufficient to show a creditable reduction in operating expense, even if there was also deducted from it any additional labor and the fixed charges on a complete equipment of special appliances for the burning of the lower-grade fuel.

Based upon the cost of the coal alone, the rate of evaporation must vary inversely as the cost in order to secure equivalent results in water evaporated per unit of cost. This principle has been carried out in the calculations of Table No. 54. Thus, if coal at $3.50 per ton has an evaporative power of unity, it will be necessary for one pound of coal at $5.00 per ton to evaporate 1.43 times as much water to produce the same commercial result; or if coal costing $1.50 per ton is substituted for coal costing $4.00 per ton, the cost per pound of water evaporated will be the same if the latter fuel evaporates 0.38 as much as the former.

Influence of Mechanical Draft.—Intense draft is one of the most important factors in the utilization of cheap fuels; hence the value of mechanical draft, by which this intensity can be most economically produced. Its influence will be discussed later at length, but it is here proper to present illustrations of the increased economy in fuel resulting from its use.

At the United States Cotton Company's mills, at Central Falls, R. I., is a boiler-plant consisting of three Babcock & Wilcox boilers of 335 horse-power each,—a total of 1005 horse-power. The draft is furnished by a 90-inch Sturtevant blower, which forces the air to the ash-pits, and whose speed is automatically regulated by a special device, so that the volume of air and intensity of draft are continually changing to suit the varied conditions and requirements of the fire.

EFFICIENCY OF FUELS.

TABLE No. 54.
RATES OF EVAPORATION FOR EQUIVALENT COST OF COAL.

Cost of Coal per Ton	Cost of Coal per Ton.						
	$0.50	$1.00	$1.50	$2.00	$2.50	$3.00	$3.50
$0.50	1.00	2.00	3.00	4.00	5.00	6.00	7.00
1.00	0.50	1.00	1.50	2.00	2.50	3.00	3.50
1.50	0.33	0.67	1.00	1.33	1.67	2.00	2.33
2.00	0.25	0.50	0.75	1.00	1.25	1.50	1.75
2.50	0.20	0.40	0.60	0.80	1.00	1.20	1.40
3.00	0.17	0.33	0.50	0.67	0.84	1.00	1.17
3.50	0.14	0.29	0.43	0.57	0.71	0.86	1.00
4.00	0.13	0.25	0.38	0.50	0.63	0.75	0.88
4.50	0.11	0.22	0.33	0.45	0.56	0.67	0.78
5.00	0.10	0.20	0.30	0.40	0.50	0.60	0.70
5.50	0.09	0.18	0.28	0.36	0.45	0.55	0.64
6.00	0.08	0.17	0.25	0.33	0.42	0.50	0.58
6.50	0.07	0.15	0.23	0.31	0.39	0.46	0.54
7.00	0.07	0.14	0.21	0.29	0.36	0.43	0.50

Cost of Coal per Ton.	$4.00	$4.50	$5.00	$5.50	$6.00	$6.50	$7.00
$0.50	8.00	9.00	10.00	11.00	12.00	13.00	14.00
1.00	4.00	4.50	5.00	5.50	6.00	6.50	7.00
1.50	2.67	3.00	3.33	3.67	4.00	4.33	4.67
2.00	2.00	2.25	2.50	2.75	3.00	3.25	3.50
2.50	1.60	1.80	2.00	2.20	2.40	2.60	2.80
3.00	1.33	1.50	1.67	1.83	2.00	2.17	2.33
3.50	1.14	1.28	1.43	1.57	1.71	1.86	2.00
4.00	1.00	1.13	1.25	1.38	1.50	1.63	1.75
4.50	0.89	1.00	1.11	1.22	1.33	1.44	1.55
5.00	0.80	0.90	1.00	1.10	1.20	1.30	1.40
5.50	0.73	0.82	0.91	1.00	1.09	1.18	1.27
6.00	0.67	0.75	0.83	0.92	1.00	1.08	1.17
6.50	0.62	0.69	0.77	0.85	0.92	1.00	1.08
7.00	0.57	0.64	0.71	0.79	0.86	0.93	1.00

Before this mechanical-draft plant was put into operation the fuel employed was George's Creek Cumberland coal, costing $4.00 per ton of 2200 pounds, delivered at the boiler-room. After the fan with automatic control was put in use, the quality and price of the fuel was reduced to a mixture of about 70 to 75 per cent of No. 2 buckwheat, 20 to 25 per cent of yard screenings, and 5 to 10 per cent of Cumberland, costing $2.62 per ton. This change reduced the fuel cost per indicated horse-power of the engine to only $5.80 per year. The annual saving in cost of fuel was about $6500.

Another illustration is that of the Hotel Iroquois at Buffalo, N. Y. The boiler-plant consists of four Babcock & Wilcox boilers aggregating 328 horse-power. The record of one year's operation with chimney-draft is presented in Table No. 55. The succeeding year the plant was operated under

TABLE No. 55.

RESULTS OF OPERATION OF BOILER-PLANT AT HOTEL IROQUOIS, BUFFALO, N. Y., WITHOUT MECHANICAL DRAFT.

Time.	Kind of Coal.	Number of Tons.	Cost per Ton.	Total Cost of each kind of Coal.	Weight and Total Cost of Coal for Year.
Dec. 1, 1892, to Nov. 30, 1893	Hard-coal screenings	232	$1.25	$1072.45	4751.24 tons
	Hard-coal screenings	601.9	1.30		
	Soft nut	696.95	2.20		$10157.38
	Soft nut	15.04	2.25	$9084.92	
	Soft nut	1759.6	2.30		
	Soft nut	1445.75	2.40		

mechanical draft, by means of cased fan with wheel 36 inches diameter by 18 inches wide, driven by a direct-connected engine, the speed of which was automatically regulated. The maximum ash-pit pressure was $1\tfrac{1}{4}$ inches of water, corresponding to a maximum fan-speed of 655 revolutions per minute. The results of one year's run under these conditions are presented in Table No. 56, from which it is evident that although

TABLE No. 56.

RESULTS OF OPERATION OF BOILER-PLANT AT HOTEL IROQUOIS, BUFFALO, N. Y., WITH MECHANICAL DRAFT.

Time.	Kind of Coal.	Number of Tons.	Cost per Ton.	Total Cost of each kind of Coal.	Weight and Total Cost of Coal for Year.
Dec. 1, 1893, to Nov. 30, 1894	Hard-coal screenings	1299.95	$1.30	$5356.24	5013 tons
	Hard-coal screenings	2610.8	1.40		
	Hard nut	3.02	3.50		$7680.93
	Soft nut	843.03	2.10	$2333.69	
	Soft nut	255.9	2.20		

the total consumption of the lower-grade coal—the burning of which was made possible by mechanical draft—was larger than that of the higher-grade coal in the preceding year, there was an actual reduction of about 25 per cent in the total cost. With chimney-draft four boilers were needed to develop the necessary amount of steam, but after the fan was introduced three boilers proved to be sufficient notwithstanding the fact that the average load during the second year was 30 horsepower greater.

The savings indicated in these cases, although the result of introducing mechanical draft, are directly attributable to the intensity of draft produced thereby and the consequent ability to burn a grade of coal which was impossible with the less intense chimney-draft.

Prevention of Smoke.—The tendency of a coal to produce smoke increases with the volatile combustible matter which enters into its composition. Pure carbon and coke are smokeless, and the best anthracite coal is practically so, but the bituminous coals are as a rule distinguished for their smoke-producing qualities. As has already been shown, the actual amount of unconsumed carbon passing away from a well-constructed boiler in the form of smoke seldom, if ever, exceeds one per cent of the total amount of carbon in the coal. Nevertheless, because of the visible effect of even such a small amount, the " smoke nuisance " is widespread and of serious consequence in some localities. Many have been the devices presented and applied for overcoming this evil, although but few have met with ultimate success.

As smoke is a result of incomplete combustion, its prevention must be sought through the provision of an ample supply of air, with sufficient intensity of draft and the maintenance of a high temperature of the fuel-bed. Unless the proposed preventive device meets these requirements it has little hope of success.

The contrivances which have been applied for the purpose may be broadly classed as follows:

I. Mechanical or forced draft.

II. Arrangements for admission of air above the fire, under which may be included steam-jets for inducing a flow of air.

III. Fire-brick arches or checker-work, placed over the bridge-wall or near the end of the fireplace, for the purpose of mixing and heating the gases.

IV. Hollow walls for preheating air.

V. Coking arches or chambers constructed in front of the fireplace, whence the coke is pushed to the rear as the volatile matter is distilled off.

VI. Double combustion, whereby part or all of the gases are passed a second time through the fuel.

VII. Down-draft furnaces in which air is admitted above the grate and the gases pass down through it and thence to the heating-surface.

VIII. Automatic stokers.

With coals of moderate smoking qualities mechanical draft in its simplest application, by its power to furnish an adequate amount of air under a pressure sufficient to cause it to pass readily through the fuel, meets all the requirements of a smoke preventive, even when all of the air is admitted beneath the grate. With excessively smoky coals, however, a portion of the air may be admitted above the fuel. The best results are obtained when this air enters the furnace under the influence of positive means, in a series of jets, by which it is forced to commingle with the gases as they rise from the bed of fuel. Owing to the tendency of cold air thus admitted to chill the fire and actually increase the amount of smoke, it is desirable, when rapid combustion takes place, that the air be preheated to a considerable degree.

The steam-jet has also been employed to induce a flow of air which, mixed with steam and thus heated, is forced either beneath or above the fire, as the case may be. While the admixture of steam incident to this method of draft-production lessens the tendency to clinker, nevertheless its cost of operation is much greater than that of the fan-blower, as will be shown later. Evidently all of the steam thus admitted must be raised to the temperature of the escaping gases, thereby reducing the efficiency of the fuel.

Hoadley * made clear the inefficiency of preheating devices where the draft is not sufficiently strong to cause the necessary movement of air, and the inadequacy of passages in brick settings as means of heating the air prior to its admission above the fire. In the boiler under test these passages were formed in the brick setting of the back and sides, in the rear of the bridge-wall, and communicated with openings at the bridge-wall and in the side walls of the furnace. Sliding dampers were provided to regulate the admission of air; but, as stated by Hoadley, " careful and repeated experiments and observations proved that these dampers could never be opened without checking the draft through the fuel and lowering the temperature of the fire; and it is not impossible that a very slight leakage through the closed dampers may have lowered the efficiency of the boilers.

As regards the arrangement " for heating air or ' superheating it ' (whatever superheating may be supposed to mean when applied to a permanent gas), . . . no good was ever found to result from this system of flues; indeed it is doubtful if any considerable quantity of air ever passed through the flues at all, although some must have flowed in when the dampers were opened, since the resistance of the open flue, circuitous as it was, could hardly have been so great as that of the coal on the fire-grates."

With a temperature of about 2000° immediately above the fire in the ordinary furnace, it is evident that any device which heats the air but a few degrees above the temperature of the atmosphere can have no practical effect upon the com-

---

* " Warm-blast Steam-boiler Furnace." J. C. Hoadley.

bustion, and that some more elaborate arrangement is necessary. As such devices pertain more properly to the efficiency of the boiler, they will be discussed under that head.

Firebrick arches and coking-chambers are both serviceable in preventing the formation of smoke, but require very careful management.

As the inflammable gas (CO) and the unconsumed carbon together seldom exceed two per cent of the total gases, any attempt to burn them again, as by " double combustion," is futile. Apparent success is due to the admission of additional air.

The down-draft furnace, owing to reversal of the usual direction of movement of the gases, requires a water-grate to withstand the intense heat. The practical features of such construction generally make it impossible to prevent considerable of the fuel dropping through the grate. To avoid loss from this source an auxiliary grate is usually provided, upon which this fuel may be consumed. Even with a forced fire and careless firing such an arrangement appears to insure a good smoke record. The conditions are such, however, that greater draft is required than with the ordinary type of furnace.

As a rule, automatic stokers are introduced for reasons other than the prevention of smoke, although, by their uniformity of feeding and their frequent application of the coking principle in their construction, they are capable, under favorable conditions, of giving good results. They do not, however, readily handle caking or clinkering coals, and usually require coal of the large sizes, and hence are restricted in their application.

With ordinary coal and hand-firing the prevention of smoke is largely dependent upon the fireman, irrespective of any special appliances; for these, no matter how excellent their character, are in course of time likely to be neglected. Many devices applied for this purpose, or for increasing the efficiency, have shown favorable results merely because they compelled greater attention on the part of the fireman in the care of boilers that were previously worked with marked inefficiency. Beyond certain factors, such as a sufficiency of draft, a good fireman is, after all, the most important factor in increasing efficiency and preventing smoke.

CHAPTER VI.

EFFICIENCY OF STEAM-BOILERS.

Measure of Efficiency.—The practical efficiency of a boiler and that of the fuel consumed in connection with it are interdependent. That is, the attainable efficiency of the fuel is dependent upon the design and operation of the boiler, its furnace, and accessories; while the efficiency of a given boiler is a direct measure of the amount of heat actually derived from the fuel employed as compared with its total calorific value. It has long been the custom to compare boilers upon the basis of the number of pounds of water evaporated by each per pound of fuel or combustible burned. Of course, efficiencies measured thus vary greatly in different types of boilers; but, broadly stated, it is undoubtedly true that the rate of evaporation per pound of coal from feed-water at 60° into steam of 80 pounds gauge-pressure is in general below 8 pounds. This is equivalent to 9.56 pounds from and at 212°. Indeed, 8 pounds of dry steam under the above conditions is a fair result, 8.25 pounds a good result, 8.5 pounds very good, and 9 pounds about the best usually attainable. This latter amount corresponds to 10.74 pounds from and at 212°, is equivalent to 69 per cent of the full calorific power of carbon, and is for coal of five sixths carbon a high result.

Results thus compared are, however, liable to be deceptive,—in some cases intentionally so,—because of the wide variation in the quality of coal. Thus, by means of picked coal of high calorific power, an exceptionally high evaporation may be obtained and used to advocate the merits of a given boiler. Less attention is generally given to the quality of the coal than to the amount of water evaporated per pound, and the fact is not always recognized that a poor boiler tested with good coal may actually give a greater evaporation than a good boiler with poor coal. The possibility of obtaining such results is rendered evident by Table No. 57, showing three sets of assumed but perfectly practical conditions.

TABLE No. 57.

RELATIVE EFFICIENCY OF BOILERS.

Designation of Boiler.	Heat-value of One Pound of Coal. B. T. U.	Evaporation from and at 212° per Pound of Coal. Pounds.	Efficiency. Per cent.
A	14500	7.50	50
B	11250	8.19	70
C	9500	6.86	70

If the measure of efficiency were to be based solely upon the evaporation per pound of coal, boiler B would be selected. But it is equalled in efficiency by boiler C, if the efficiency be measured by the proportion of available heat utilized, although it evaporates 1.33 pounds less water because of the poorer quality of the coal. Whether it would be commercially the more efficient of the two would depend upon the relative cost of the two kinds of coal, and that having a heat-value of 9500 B. T. U. would have to cost about 14 per cent less. Based upon water evaporated, boiler A appears more efficient than boiler C, but measured by the amount of heat absorbed the latter far exceeds the former. Although boiler A, under the given conditions, evaporates about 9 per cent more than boiler C, boiler C is the more economical in the combustion of coal by about 40 per cent.

It is evident that the evaporation from and at 212° per pound of fuel, although a convenient basis of comparison for fuels, is not properly applicable for defining the efficiency of a boiler. It is thus that the distinction is to be drawn between the efficiency of a fuel and that of the boiler in connection with which it is burned. A more accurate basis for the comparison of efficiency of different types of boilers is established when the evaporation is expressed in pounds of water evaporated per pound of combustible. But even this is somewhat affected by the proportion of ash and elementary water in the coal.

The ideal basis of comparison is to be sought in the ratio found by experiment to exist between the total effective heat-value of the coal, as determined by means of a calorimeter and analysis, and that rendered evident in the steam generated, which may be thus expressed:

$$\text{Efficiency} = \frac{\text{Heat-units usefully applied.}}{\text{Heat-units supplied to furnace.}}$$

Although ideally the correct method of comparing the efficiency of boilers, and likewise of fuels, when proper allowances can be made for boiler differences, it is to a certain extent open to criticism because of the difficulty of determining experimentally from a small specimen the exact heating-value of the entire quantity of coal. Even with this basis of comparison, relative efficiency-tests should be conducted under identical conditions so far as they may be obtainable. The complete results can be best presented in the form of a heat balance, as follows:

HEAT BALANCE.

Dr.		Cr.

To heat
  from coal;
  from air;
  from feed-water.
	By heat
	  in dry steam;
	  in moisture and water mechanically suspended
	    in steam;
	  in dry flue-gases;
	  in moisture in coal;
	  in water resulting from combustion,
	  in vapor in air,
	  lost through incomplete combustion to CO;
	  in ashes;
	  lost by radiation and otherwise unaccounted for.
		at temperature of flue-gases;

As is the case with fuel, so is it with a boiler: the efficiency must be considered commercially. For this reason a limit is reached considerably short of 100 per cent, beyond which the loss in interest, depreciation, and other fixed charges exceeds the gain from decreased cost of fuel, per unit of evaporation, resulting from the given improvement.

In illustration of the difference in efficiency of different types of boilers, Table No. 58 is presented. This covers the

TABLE NO. 58.

EFFICIENCY OF DIFFERENT TYPES OF BOILERS.

Kind of Boiler.	Number of Trials.	Efficiency in Per Cent.		
		Maximum.	Minimum.	Average.
Small vertical..........................	3	46.10	34.60	41.60
Large vertical.........................	3	52.30	49.00	50.90
Large vertical, improved setting.....	1	....	One trial only	67.89
Tubular boilers.......................	14	60.17	44.76	51.53
Tubular boilers, improved settings ..	34	76.38	41.94	58.87
Water-tube boilers...................	13	70.11	49.37	61.31
Water-tube boilers, improved setting	18	81.32	49.30	67.52

results of 86 tests conducted by Bryan[*] under common conditions and with ordinary fuel, principally Illinois coal. Due allowance is to be made for locality, special type of boiler, and kind of coal.

Rating of Steam-boilers.—As originally the most important purpose of the steam-boiler was to generate steam for use in a steam-engine, it became customary to express its capacity by the nominal output of the engine which it supplied. That is, its rating was expressed in horse-power (a term with which the boiler has properly nothing to do); a horse-power in each case representing the weight of steam required per hour to enable the engine to perform work continuously at the rate

---

[*] "Boiler Efficiency, Capacity, and Smokelessness, with Low-grade Fuels." Wm. H. Bryan. A paper read before the Engineers' Club of St. Louis, Oct. 21, 1896.

of 33,000 foot-pounds per minute. The first standard, fixed by Watt, and based upon the performance of the engines of his day, was one cubic foot of water (weighing about 60 pounds) evaporated per hour from 212° per horse-power.

As the efficiency of the steam-engine has been improved, the amount of steam necessary for the production of a horse-power has been gradually decreased. The attempt was made to keep pace with engine improvements by correspondingly reducing the amount of water represented by a horse-power. But the present existence of engines, in great variety of design and manner of operation, as compared with the simple type in the days of Watt, renders impossible the establishment of any definite standard of rating which shall apply to them all. This cannot be more clearly evidenced than by Table No. 59, which embodies Thurston's* estimate of the steam-consumption of the best classes of engines in common use and in good

TABLE No. 59.

STEAM PER HORSE-POWER PER HOUR FOR STEAM-ENGINES WITH DIFFERENT RATIOS OF EXPANSION.

Type of engine.	Steam-pressure above Atmosphere.	Ratio of Expansion.					
		2	3	4	5	7	10
Non-condensing	30	40	39	40	40	42	45
	45	35	34	36	36	38	40
	60	30	28	27	26	30	32
	75	28	27	26	25	27	29
	90	26	25	24	23	25	27
	105	25	24	23	22	22	21
	135	24	23	22	21	20	20
Condensing	15	30	28	28	30	35	40
	30	28	27	27	26	28	32
	45	27	26	25	24	25	27
	60	26	25	25	23	22	24
	75	26	24	24	22	21	20
	105	25	23	23	22	21	20
	135	25	23	22	21	20	19

* "A Manual of Steam-boilers." R. H. Thurston.

order. Evidently the term "horse-power," applied to the rating of steam-boilers, must, therefore, be considered as a standard of measurement rather than a direct measure of capacity.

In 1876 the committee of judges of the Centennial Exhibition to whom was intrusted the trials of the competing boilers exhibited, decided to adopt a standard rating upon conditions considered by them to represent fairly average practice. The standard unit of 30 pounds of water of 100° temperature evaporated into dry steam of 70 pounds gauge-pressure, thus adopted, has since become the almost universal standard of rating in the United States for the nominal evaporative capacity of steam-boilers, and is commonly designated as a commercial horse-power. The amount of heat required to evaporate one pound of water under these conditions is 1110.2 B. T. U., equivalent to 1.1496 units of evaporation, as previously defined. An evaporation of 30 pounds under the stated conditions is, therefore, equivalent to the development of 33,305 B. T. U. per hour, or 34.488 pounds of water evaporated from and at 212°, or in round numbers 34.5 pounds. Although even this rating is now open to criticism, because it does not represent the present standard of average steam-engine performance, nevertheless, being a more or less arbitrary standard, it is as satisfactory as any other for the mere purpose of comparing boiler capacities, and has in substance been officially adopted by the American Society of Mechanical Engineers.* Any standard of this character is, however, open to the criticism that because of the range of capacity possessed

---

* In its Code of 1898 the standard is expressed as follows:

"The unit of commercial horse-power developed by a boiler shall be taken at 34½ units of evaporation per hour; that is, 34½ pounds of water evaporated per hour from a feed-water temperature of 212 degrees Fahr. into dry steam of the same temperature. This standard is equivalent to 33,317 British thermal units per hour. It is also practically equivalent to an evaporation of 30 pounds of water from a feed-water temperature of 100 degrees Fahr. into steam of 70 pounds gauge-pressure." Transactions American Society of Mechanical Engineers, vol. xx.

by any boiler it is difficult to fix the conditions under which this capacity should be attained. Should they be those that exist when the boiler is being run easily under ordinary conditions, or should the measure of capacity be taken when the boiler is pushed to its utmost? The above-mentioned Code of 1898 specifies that "a boiler rated at any stated capacity should develop that capacity when using the best coal ordinarily sold in the market where the boiler is located, when fired by an ordinary fireman, without forcing the fires, while exhibiting good economy; and, further, that the boiler should develop at least one third more than the stated capacity when using the same fuel and operated by the same fireman, the full draft being employed and the fires being crowded; the available draft at the damper, unless otherwise understood, being not less than ½-inch water-column."

Table No. 60 is presented for the purpose of simplifying the reduction of horse-power under any given conditions to that under other conditions.

The basis is 30 pounds of water at 100° evaporated into steam of 70 pounds pressure. At any other temperature, as for instance 170°, and pressure of, say, 80 pounds, the equivalent evaporation is shown to be 31.96 pounds.

The preceding applies more directly to the rating of boilers as determined by experimental test. But it is obviously desirable that, for the purposes of designation, the capacity or rating of a given design of boiler should be at least approximately known before it is constructed. Evidently such a rating must be based upon previous experiment with similar boilers, having the same general proportions and disposition of heating-surface and operating under similar conditions. The measure of such results is made in the number of pounds of water evaporated per hour per square foot of heating-surface, or its equivalent, the number of square feet of heating-surface per commercial horse-power. In boilers of the same relative proportions of grate- to heating-surface, and of the same general arrangement of heating-surface, the rate of

EFFICIENCY OF STEAM-BOILERS. 147

TABLE No. 60.

REQUIRED HOURLY EVAPORATION PER COMMERCIAL HORSE-POWER AT VARIOUS TEMPERATURES OF FEED AND PRESSURE OF STEAM.

Temperature of Feed-water. Deg. Fahr.	Steam-pressure above the Atmosphere.																				
	0	10	20	30	40	50	60	70	80	90	100	110	120	130	140	150	160	170	180	190	200
50	29.51	29.20	29.14	29.02	28.92	28.84	28.77	28.70	28.64	28.59	28.54	28.49	28.45	28.41	28.37	28.33	28.29	28.26	28.23	28.20	28.17
60	29.77	29.55	29.40	29.28	29.18	29.09	29.02	28.95	28.89	28.84	28.79	28.74	28.69	28.65	28.61	28.57	28.54	28.51	28.48	28.45	28.42
70	30.04	29.81	29.66	29.54	29.44	29.35	29.27	29.21	29.15	29.09	29.04	28.99	28.94	28.90	28.86	28.82	28.78	28.75	28.72	28.69	28.66
80	30.31	30.08	29.93	29.80	29.70	29.61	29.53	29.46	29.40	29.34	29.29	29.24	29.19	29.15	29.11	29.07	29.03	29.00	28.97	28.94	28.91
90	30.59	30.36	30.20	30.07	29.97	29.88	29.80	29.73	29.67	29.61	29.55	29.50	29.45	29.41	29.37	29.33	29.29	29.25	29.22	29.19	29.16
100	30.88	30.64	30.47	30.34	30.24	30.15	30.07	30.00	29.93	29.87	29.82	29.77	29.72	29.67	29.63	29.59	29.55	29.51	29.48	29.45	29.42
110	31.17	30.93	30.76	30.63	30.52	30.43	30.34	30.27	30.20	30.14	30.09	30.04	29.99	29.94	29.90	29.86	29.82	29.78	29.74	29.71	29.68
120	31.46	31.22	31.05	30.91	30.80	30.71	30.63	30.55	30.48	30.42	30.36	30.31	30.26	30.21	30.17	30.13	30.09	30.05	30.01	29.98	29.95
130	31.76	31.52	31.34	31.20	31.09	30.99	30.91	30.83	30.76	30.70	30.65	30.59	30.54	30.49	30.45	30.41	30.37	30.33	30.29	30.25	30.22
140	32.07	31.82	31.64	31.50	31.38	31.29	31.20	31.12	31.05	30.99	30.93	30.88	30.83	30.78	30.73	30.69	30.65	30.61	30.57	30.53	30.50
150	32.39	32.12	31.94	31.80	31.68	31.58	31.50	31.42	31.35	31.28	31.22	31.17	31.12	31.07	31.02	30.97	30.93	30.89	30.85	30.81	30.78
160	32.71	32.44	32.26	32.11	31.99	31.89	31.80	31.72	31.65	31.58	31.52	31.46	31.41	31.36	31.31	31.27	31.23	31.19	31.15	31.11	31.08
170	33.03	32.76	32.58	32.43	32.31	32.20	32.11	32.03	31.96	31.89	31.83	31.77	31.71	31.66	31.61	31.56	31.52	31.48	31.44	31.40	31.37
180	33.37	33.09	32.90	32.75	32.63	32.52	32.43	32.34	32.27	32.20	32.14	32.08	32.02	31.97	31.92	31.87	31.83	31.79	31.75	31.71	31.67
190	33.71	33.43	33.23	33.08	32.95	32.84	32.75	32.66	32.59	32.50	32.45	32.39	32.33	32.28	32.23	32.18	32.14	32.10	32.06	32.02	31.98
200	34.06	33.77	33.57	33.41	33.28	33.17	33.08	32.99	32.91	32.84	32.77	32.71	32.65	32.60	32.55	32.50	32.45	32.41	32.37	32.33	32.29
212	34.49	34.18	33.98	33.80	33.69	33.58	33.48	33.39	33.31	33.24	33.17	33.11	33.05	32.99	32.94	32.89	32.84	32.80	32.76	32.72	32.68

evaporation is fairly constant for ordinary draft conditions. That is, the total evaporation is practically proportional to the heating-surface. The area of this surface, therefore, forms a ready means of comparison of capacity. But the location and character of the heating-surface largely determines the rate of evaporation, thus restricting direct comparison to boilers of the same type under the same conditions. A fair average for boilers of the common horizontal return tubular type, with ordinary natural draft, is an evaporation of 3 pounds of water from and at 212° per square foot of heating-surface. On the basis of 34.5 pounds per horse-power, under these conditions, this is equivalent to one horse-power for each 11.5 square feet of heating-surface. It is true, however, that the horse-power of boilers of this type is still quite generally figured on a basis of 15 square feet. The number of square feet of heating-surface per horse-power, for other rates of evaporation, is presented in Table No. 61.

TABLE No. 61.

SQUARE FEET OF HEATING-SURFACE PER HORSE-POWER.

Pounds of water evaporated from and at 212° per square foot of heating-surface per hour,	2.0	2.5	3.0	3.5	4.0	5.0	6.0	7.0	8.0	9.0	10.
Square feet of heating-surface required per horse-power,	17.3	13.8	11.5	9.8	8.6	6.8	5.8	4.9	4.3	3.8	3.5

Under mechanical draft the rate of evaporation per unit of heating-surface may be greatly increased. In fact, it has been generally applied, particularly in the marine service, with a view to lessening the dimensions of the boiler for the same output. This may be rendered possible by the fire, owing to its increased temperature, being able to transmit more heat per unit of surface, thereby making the boiler, surface for surface, more efficient than it would be if a lower fire-temperature were used. It is evident that if the boiler-surfaces are more efficient, less heat will be wasted up the chimney.

The horse-power of various types of boilers, based upon

the area of heating-surface shown by experiment to be necessary for the evaporation of 34.5 pounds of water from and at 212° per hour, is usually accepted as the nominal rating; owing to slight differences even between boilers of the same type, there is some latitude in this basis of measurement. The present general range of proportions is indicated in Table No. 62.

TABLE NO. 62.

RELATION OF HORSE-POWER AND HEATING-SURFACE IN DIFFERENT TYPES OF BOILERS.

Type of Boilers	Rate of Combustion.	Sq. Ft. of Heating-surface per Sq. Ft. of Grate.	Average Equivalent Evaporation.	Sq. Ft. of Grate per Boiler H. P.	Heating-surface per Boiler H. P.
Lancashire	8 to 10	25 to 30	8 to 10	0.36	7.0
Cylindrical multitubular	8 to 15	35 to 40	9 to 10.5	0.30	11.5
Vertical, Manning	10 to 20	*48 to 16	9 to 10.5	0.23	11.1
Locomotive	50 to 120 average 75	60 to 70	6.7 to 8.5	0.07	4.5
Locomotive type, stationary	8 to 15	40 to 45	9 to 10.5	0.30	12.6
Scotch marine	35 to 45	30	7 to 9	0.11	3.3
Water-tube, with cylinder or drum.	9 to 15	35 to 45	9 to 10.5	0.28	11.0
Water-tube, with separator	15 to 67 average 20	30 to 40	7 to 9	0.22	7.3

* 48 heating-surface, 16 superheating-surface.

"This table has been compiled from a large number of examples, and may be taken to represent current good practice. The last two columns, giving the grate-surface and heating-surface, have been compiled for 34 5 pounds of water evaporated per hour from and at 212° F."*

Radiation and Convection of Furnace-heat.—The ideal temperature of combustion of a coal consisting of 80 per cent carbon has already been shown by Table No. 18 to be 4645° when the chemically necessary amount of air is supplied. If this supply be doubled, the temperature is reduced to 2557°; in each case the temperature being measured above that of the air supplied to the furnace. These ideal temperatures pertain to the heart of the fire and can only exist in the fur-

* "Steam-boilers." Peabody and Miller.

nace-chamber if the fire is properly inclosed with radiating surface.

The ordinary boiler-plate, with the hot gases on one side and water on the other, presents a very different condition, for it becomes a greedy absorber of heat, both radiant and convected. The air and gases, being poor conductors of heat, and absorbing but a very slight amount of the radiant heat, have only the power to increase the temperature of the surface with which they come in contact by the process of convection or carrying, which may be defined as "a transfer and diffusion of the state of heat in a fluid mass by means of the motion of the particles of that mass." *

The heat that is radiated from the fire is but feebly reciprocated from the plate-surfaces of the boiler, since the plate is maintained at a temperature not much higher than that of the water inside. Under such conditions the heat which is radiated from the fuel upon the grate, together with that which is communicated by convection from the heated gases, is rapidly absorbed and carried off. It is, therefore, impossible to maintain in the furnace a temperature even near the maximum temperature of combustion. The radiation from the fuel, taking place, as it does, in straight lines, is thereby restricted in its effect to the grates, the walls of the furnace-chamber, and the exposed portion of the boiler. The heat not thus lost is carried along by the products of combustion.

The exact proportional relation between the radiant and convected heat is difficult of determination, but Peclét assumed them to be equal as they leave the upper surface of the fuel. Upon this assumption and the formula derived by Dulong and Petit, the relation of radiant and convected heat has been calculated by Clark † for different rates of combustion. The conditions imposed are: complete combustion, no excess of air, and a coal having a heat-value of 14,700 B. T. U. The results are given in Table No. 63, with the average rela-

---
* "A Manual of the Steam-engine." W. J. M. Rankine.
† "The Steam-engine." D. K. Clark.

tive proportions existing between the two means of dissipation of heat.

TABLE No. 63.

TEMPERATURE AND HEAT OF A COAL FIRE IN A STATE OF INCANDESCENCE.

Coal Consumed per Square Foot of Grate per Hour.	Temperature of Surface of Boiler-plate.	Temperature of Surface of Fire.	Approximately Calculated Distribution of the Heat of Combustion.					
			By Radiation.		By Convection of Gases.		Sum of Radiation and Convection.	
Pounds.	Deg. Fahr.	Deg. Fahr.	B. T. U.	Per cent of Sum.	B. T. U.	Per cent of Sum.	B. T. U.	
5	350°	1400°	53960	74	19160	26	73120	
10	350	1550	102500	70	43510	30	146010	
20	350	1705	198400	67	96080	33	294480	
40	350	1857	378650	64	209850	36	588500	
80	350	2009	721800	61	455400	39	1177200	
120	350	2097	1049000	59	714050	41	1763050	

Distribution of the Heat of Combustion.—It being the function of a boiler to utilize as much as possible of the heat generated by the combustion of the fuel in the furnace provided for the purpose, it is to be expected that the gases will be gradually cooled as they approach the chimney. The greater the decrease, and, other things equal, the lower the final temperature, the higher the efficiency. The temperature of the gases at different points in their passage across the heating-surface of the boiler obviously depends upon the character, extent, and temperature of that surface, and the initial temperature and velocity of the gases.

The absorption of practically all of the radiant and part of the convected heat by the surfaces exposed to the fire immediately lowers the temperature of the gases so that as they pass onward to the other portions of the heating-surface their heating-power is much decreased. The existence of this condition is rendered evident by the results of tests reported by Havrez.* These tests were conducted upon a special boiler of the locomotive type, the barrel or cylindrical portion of which was divided into four sections of equal length, and so arranged that the evaporation in each could be ascertained separately. The general proportions of this boiler, together with the results, are presented in Table No. 64, from which it is evident that the surface of the fire-box was about three times as efficient as that of the first section, and increasingly more for the other sections respectively.

TABLE No. 64.

EVAPORATION IN DIFFERENT SECTIONS OF EXPERIMENTAL BOILER.

Portion of Boiler.	Square Feet of Heating-surface.	Pounds of Water Evaporated per Hour per Sq. Ft. of Heating-surface.	
		With Coke for Fuel.	With Briquettes for Fuel.
Fire-box	76.43	24.5	36.9
First section	179	8.72	11.44
Second second	179	4.42	5.72
Third section	179	2.52	3.52
Fourth section	179	1.68	2.31

The variation in the rate of absorption of heat from the gases, which takes place in their passage through the tubes of a marine boiler, is clearly shown by the results of tests conducted under the supervision of Engineer-in-Chief Durston of the British Navy, as presented in Table No. 65. These results cover eight sets of records of temperatures taken by a Le Chatelier thermoelectric pyrometer inserted to various distances in the tubes. The boiler was being worked at its normal capacity, the rate of consumption of coal being about 17 pounds per square foot of grate. The temperature exist-

* Proceedings of Institute of Civil Engineers.

ing in the combustion-chamber was 1644°, and that just inside the tube 1550°.

TABLE No. 65.

TEMPERATURES IN TUBES OF MARINE BOILER.

Location.	Temperature.
1 inch from combustion-chamber	1466°
2 inches from combustion-chamber	1426
3 " " "	1405
4 " " "	1412
5 " " "	1398
6 " " "	1406
7 " " "	1400
8 " " "	1410
1 ft. 2 in. from combustion-chamber	1368
1 ft. 8 in. " "	1295
2 ft. 8 in. " "	1198
3 ft. 8 in. " "	1106
4 ft. 8 in. " "	1015
5 ft. 8 in. " "	926
6 ft. 8 in. " "	887
In smoke-box	782

The temperatures existing under usual conditions, in connection with an ordinary horizontal return tubular boiler, were very carefully ascertained by Hoadley,* and are presented in Table No. 66. The coal burned consisted of 82 per cent carbon completely consumed to carbonic acid, the steam-pressure was about 45 pounds above the atmosphere, and the

TABLE No. 66.

TEMPERATURES IN CONNECTION WITH HORIZONTAL RETURN TUBULAR BOILER.

Location at which Temperature was Taken.	Temperature.
In heart of fire	2426°
At bridge-wall	1341
In smoke-box	368
Air admitted to furnace	78
Steam and water in boiler	292
Gases escaping to chimney	368

* "Warm-blast Steam-boiler Furnace." J. C. Hoadley.

air-supply per pound of fuel was 21.28 pounds. The actual temperature shown to exist, as compared with that which the coal should be ideally capable of producing, with no excess of air, as previously indicated, is particularly to be noted.

Disposition of Heat in Steam-boilers.—The theoretical heat-losses incident to the combustion of a given amount of coal have already been considered in the preceding chapter. These indicate clearly the disposition of all the heat generated, with the exception of that lost through the brickwork. This loss is evidently variable and uncertain, but under ordinary conditions is between 4 and 20 per cent. Hoadley[*] conducted a series of experiments upon the setting of a horizontal return tubular boiler, by inserting thermometers to different depths in holes in the walls at 5 feet above the floor,—opposite to the body of the boiler and midway between the bridge-wall and the pier. The maximum and minimum temperatures found between 8.30 A.M. and 4.30 P.M., at different depths, upon one day of the test, are presented in Table No. 67. Of course the temperature at the outer surface of the setting would be considerably less than that 4 inches inward, owing to the rapid radiation at the surface.

TABLE No. 67.

TEMPERATURES OF BRICK SETTING OF HORIZONTAL RETURN TUBULAR BOILER.

Place of Observation.	Minimum Temperature.	Maximum Temperature.
4 inches from outside of wall............	140°	182°
16 " " " " "	285	353
28 " " " " "	297	460

The disposition of the heat generated in the furnace of a steam-boiler, as determined by several investigators for different types of boilers, is indicated in Table No. 68. The boiler experimented on by Bunte was of ordinary character; that

---

[*] " Warm-blast Steam-boiler Furnace." J. C. Hoadley.

EFFICIENCY OF STEAM-BOILERS.   155

TABLE No. 68.
DISPOSITION OF HEAT IN STEAM-BOILERS.

Disposition of Heat.	Authority.					
	Bunte.	Scheurer and Meunier.		Donkin and Kennedy.		Hoadley.
		A	B	C	D	E	
Waste in flue-gases including evaporation of moisture in coal and heating vapor in air when these losses are not separately given.	18.6	5.5	14.8	9.4	22.5	6.5	5.04
Evaporating-moisture in coal.................	3.5	2.5	6.1	0.1	0.1	0.0	1.55
Heating vapor in air...	......	......	......	......	......	......	0.18
Imperfect combustion.......................	8.0	6.0	......	12.7	0.0	0.0	1.44
Clinker and ash.............................	4.1	......	......	0.1	0.2	0.0	......
Radiation and heat not otherwise accounted for	7.6	23.5	13.4	13.9	11.0	15.0	4.00
Heating and evaporation of water...........	58.2	61.0	65.7	63.8	66.2	78.5	87.79

designated A and reported upon by Scheurer and Meunier was of the French type, having three heaters with six feed-heater tubes at one side; while the results given under B are the average of experiments upon four different types of boilers. Boiler C, experimented upon by Donkin and Kennedy, was of vertical tubular construction with internal fire-box; boiler D was of the locomotive pattern; while E was of the "elephant" type, provided with Green's economizer. The boiler tested by Hoadley was of the ordinary multitubular form, provided with a special air-heating arrangement which lowered the temperature of the flue-gases about 213°, and raised that of the air supplied to the furnace about 300°. These abstractors account for the high efficiency secured, which might have been closely approached by boiler E had it not been for the excessive loss through the brickwork.

All figures are given in percentages of the total amount of heat accounted for. The percentage of heat which is indicated to have been disposed of in heating and evaporating the water is a direct measure of the efficiency of the boiler.

Sources of Efficiency.—With a given fuel, and otherwise identical conditions, the efficiency of a boiler is most largely dependent upon the relation of its grate-area to its heating-surface, and upon the rate of combustion of the fuel. Under the same conditions of boiler and fuel, the greater the quantity consumed per hour the greater is the amount of water evaporated per hour. But at the same time the tendency is that the quantity of water evaporated per pound of fuel will decrease because of the probable higher temperature of the escaping gases. This loss, if it occurs, can only be diminished by increasing the heating-surface either in the boiler or in a separate heater, while the decreased draft, due to lowered temperature, is easily made good by mechanical means. It is to be carefully noted that all this relates to a condition of increased boiler-capacity, resulting from a greater coal-consumption. But when the capacity and coal-consumption are maintained practically constant, a higher efficiency per pound of fuel may actually be obtained by a moderate reduction of grate-area, whereby the surface-ratio is increased, with a corresponding increase in the rate of combustion per square foot of grate. This matter of the relation between the surface-ratio and the rate of combustion and its economical influence, as indicated in the evaporation per unit of fuel, will be discussed at considerable length in the succeeding chapter.

The quality of the fuel must to a certain extent enter into the problem of boiler-design to secure the highest efficiency. The desirable features to give the best results with low-grade fuels are concisely stated by Bryan[*] to be:

" A. Ample draft; 1 inch of water or even more. Good results cannot be secured with drafts less than one half inch. Good draft and thick beds of fuel permit high fire-box temperatures, which we have found absolutely necessary.

" B. Large ratio of heating- to grate-surface, so that while burning coal at a high rate per square foot of grate per hour, there is sufficient heating-surface to reduce the temperature of the flue-gases to 450° F. or less.

" C. The combustion-chamber should, if possible, be

---

[*] "Boiler Efficiency, Capacity, and Smokelessness, with Low-grade Fuels." Wm. H. Bryan. A paper read before the Engineers' Club of St. Louis, Oct. 21, 1896.

separate from the heating-surfaces, so as to avoid their cooling effect. It should be quite deep—30 inches or more."

As indicating the importance of the draft, and the desirability of mechanical means for creating the same, Bryan states that: "To secure the very highest results, the gases, after leaving the boiler-heating surfaces at not exceeding 500°, should be passed through feed-water economizers and thence through air-heaters. The feed-water, leaving the ordinary exhaust-heater at a little above 200° F., may be raised to over 300° in the economizer, and the heated gases reduced to 250° or less. This reduction in temperature, of course, destroys the usefulness of these gases as draft-producers, unless the chimney is very tall. The draft, however, can be better produced by exhaust-fans, which draw the air through and out of the furnace and economizer, and discharge the gases at such a height above the roof that they will not be objectionable; thus doing away entirely with the necessity for high chimneys. Still better economy may be secured by placing air-heaters in the smoke-flue, beyond the fan or between it and the economizer. Through this the air, entering the ash-pit for purposes of combustion, may be drawn, so that the heated gases are finally discharged at a temperature but little above that of the atmosphere. The speed of the fan may be controlled by an automatic regulator, which increases the speed of the fan-engine as the steam-pressure drops, and reduces it as the pressure increases; thus performing all the functions of an automatic damper-regulator. This plan is not experimental or untried, but has already been adopted in numerous large plants."

The influence of the relation between area of grate and of heating-surface was very carefully investigated by Clark[*] in the case of a locomotive-boiler using coke. From these tests he deduced—

"*First.* That, assuming throughout a constant efficiency

---

[*] "The Steam-engine." D. K. Clark.

of the fuel or proportion of water evaporated to the fuel, the evaporative performance of a locomotive-boiler, or the quantity of water which it is capable of evaporating per hour, *decreases* directly as the grate-area is increased; that is to say, the larger the grate the smaller is the evaporation of water when the efficiency of the fuel is the same, even with the same heating-surface.

"*Second.* That the evaporative performance *increases* directly as the square of the heating-surface, with the same area of grate and efficiency of fuel.

"*Third.* The necessary heating-surface *increases* directly as the square root of the performance; that is to say, for example, for four times the performance, with the same efficiency, twice the heating-surface only is required.

"*Fourth.* The necessary heating-surface *increases* directly as the square root of the grate, with the same efficiency; that is to say, for instance, if the grate be enlarged to four times its first area, twice the heating-surface would be required, and would be sufficient to evaporate the same quantity of water per hour with the same efficiency of fuel."

The relation between the area of grate and of heating-surface, which has already been expressed as the "surface-ratio," may be thus represented:

$$\text{Surface ratio} = \frac{\text{Area of heating-surface}}{\text{Area of grate-surface}}.$$

This ratio naturally varies according to the type of boiler, the general practice being about as indicated in Table No. 69. Somewhat different figures are given by Peabody and Miller, see Table No. 62.

As it has already been shown that when the total coal-consumption is increased the heating-surface must also be increased to maintain the same efficiency, so the converse must be evident: that with a given rate of combustion it is not economical to increase the area of heating-surface beyond certain limits. These limits must of necessity be determined by experiment.

TABLE No. 69.

SURFACE-RATIOS OF STEAM-BOILERS.

Type of Boiler.	Area of Heating-surface when Grate Area = 1.
Marine return tubular	25 to 38
Lancashire	26 to 33
Cornish	25 to 40
Modified locomotive type	30 to 34
Horizontal return tubular	30 to 50
Water-tube	35 to 65
Horizontal internally fired multitubular	25 to 45
Locomotive	60 to 90
Plain cylinder	10 to 15

Clark* summarized his deductions from a large number of tests of boilers of different types in the following formulæ:

Stationary boilers........ $w = 0.0222r^2 + 9.56c$.
Marine boilers........ $w = 0.016r^2 + 10.25c$.
Portable-engine boilers........ $w = 0.008r^2 + 8.6c$.
Locomotive-boilers (coal-burning)... $w = 0.009r^2 + 9.7c$.
Locomotive-boilers (coke-burning).. $w = 0.0178r^2 + 7.94c$.

In which $w =$ weight of water in pounds per square foot of grate per hour;

$c =$ pounds of fuel per square foot of grate per hour;

$r =$ ratio of heating- to grate-surface.

The water is taken as evaporated from and at 212°.

The ratio of grate-surface to heating-surface being one of the factors, it is evident that by means of these formulæ the evaporation per square foot of heating-surface may also be obtained. There are minimum rates of consumption of fuel below which these formulæ are not applicable. The limit varies for each kind of boiler and with the surface-ratio. It

---

* "The Steam-engine." D. K. Clark.

is imposed by the fact that the maximum evaporative power of fuel is a fixed quantity, and is naturally at that point where the reduction of the rate of combustion for a given ratio procures the absorption into the boiler of the whole of the proportion of the heat which is available for evaporation. In the combustion of good coal the limit of evaporative efficiency may be taken as measured by 12.5 pounds of water from and at 212°. Table No. 70, based upon these formulæ, presents

TABLE No. 70.

EVAPORATIVE PERFORMANCE OF STEAM-BOILERS FOR INCREASING RATES OF COMBUSTION AND DIFFERENT SURFACE RATIOS AND BEST COAL AND COKE.

Kind of Boiler.	Water from and at 212° per Hour.	Fuel per Square Foot of Grate per Hour in Pounds.					
		5	10	15	20	30	40	50
For Best Coal. Surface Ratio = 30. Stationary....	Per square foot of grate....	62.5*	116	163	211	307	402	498
Stationary....	Per pound of coal.........	12.5	11.56	10.89	10.56	10.23	10.06	9.96
Marine......	Per square foot of grate....	62.5*	117	168	219	322	424	527
Marine.......	Per pound of coal.........	12.5	11.69	11.25	10.95	10.69	10.61	10.54
Portable.....	Per square foot of grate....	50.0	93	136	179	265	351	437
Portable.....	Per pound of coal.........,	10.0	9.3	9.01	8.95	8.83	8.77	8.74
Locomotive..	Per square foot of grate....	57.0	105	154	202	299	396	493
Locomotive..	Per pound of coal.........	11.4	10.5	10.26	10.10	9.97	9.90	9.86
Surface Ratio = 50. Stationary....	Per square foot of grate....	62.5*	125*	187.5*	247	342	438	534
Stationary....	Per pound of coal.........	12.5	12.5	12.5	12.33	11.41	10.95	10.67
Marine.......	Per square foot of grate....	62.5*	125*	187.5*	245	348	450	552
Marine.......	Per pound of coal........	12.5	12.5	12.5	12.25	11.58	11.25	11.05
Portable....	Per square foot of grate....	62.5*	106	149	192	278	364	450
Portable.....	Per pound of coal.........	12.5	10.6	9.93	9.6	9.27	9.10	9.00
Locomotive..	Per square foot of grate....	62.5*	120	168	217	314	411	508
Locomotive..	Per pound of coal.........	12.5	11.95	11.20	10.85	10.45	10 26	10.15

* These quantities fall below the scope of the formulæ for the water, as explained in the text.

the effect of increasing rates of combustion with different surface-ratios. While for a given boiler and surface-ratio this table indicates that the evaporative efficiency decreases as the rate of combustion is increased, it is to be noted that the capacity of the boiler is increased also, and that by a proper application of more heating-surface the efficiency may be maintained. In other words, a boiler and its appurtenances should be designed for a given rate of combustion. A reasonably high rate of combustion is not an absolute indication of low efficiency; but, as will be shown later, may be, with a proper surface-ratio, one of the important factors in attaining economy in boiler-practice.

By the same process of reasoning employed in the previous chapter, in the discussion of the influence of the air-supply, it is evident that the higher temperature of the escaping gases resulting from an increased total coal-consumption is due to the increased supply of air and the higher velocity which, therefore, ensues. Although this higher temperature is but the result of more rapid combustion, it is at the same time an absolute necessity when a chimney alone is depended upon to create the draft. This is because the draft required for the increased combustion and air-supply can only be secured, in the case of a chimney, by raising the temperature of the chimney-gases. Under these conditions any attempt to reduce this final temperature by the addition of heating-surface must of necessity tend to reduce the rate of combustion. Furthermore, the influence upon the draft of such additional surface will be twofold: it will be reduced both because of the lower temperature in the chimney, and because of the increased resistance due to the extended surface. The rate of combustion can only be maintained by supplementing the draft, which may be readily done by introducing a fan either for forcing in the air or withdrawing the gases.

But, with sufficient draft, higher efficiency may evidently be secured by increasing the heating-surface, with the limitation that the surface shall not be so great as to cool the gases too near to the temperature of the steam; for it is probable that there can be no active transmission of heat from the gases without to the water within a boiler, with less than 75° difference of temperature. Nevertheless, one of the vital principles underlying the attainment of economy in the generation of steam is a low temperature of the escaping gases.

The loss of efficiency due to the temperature of the escaping gases has already been presented in Table No. 48. The line shown in Fig. 10, based on the figures given in that table, serves to make clear the rate and proportion of this loss.

The opportunity for securing more economic results by reducing the temperature of the flue-gases is well evidenced in the results of seventeen independent boiler-tests by Donkin and Kennedy. They found the heat lost up the stack, where no economizer was used, to range between 9.4 per cent and 31.8 per cent of the total heat of combustion, the average being 20.3 per cent. It is, therefore, evident that in this direction lies one of the greatest opportunities for increasing boiler-efficiency. Although additional surface may be obtained by reducing the size of the tubes and increasing their number, or by ribbing them, or introducing retarders, it is usually customary to abstract the surplus heat from the gases by some means in a sense independent of the boiler. It may then take the form of a feed-water heater, otherwise known as an economizer, or the form of a device for abstracting the heat from the gases and transferring it to the air supplied to the fuel, or both. The results obtained by either of these methods have, in the case of chimney-draft, usually been restricted by the cost of the excessively high chimney necessary to produce the requisite draft with the decreased temperature and increased resistance. Mechanical draft, however, obviates this difficulty, and makes possible the attainment of much lower final temperatures of the flue-gases with a corresponding increase of efficiency.

Flue Feed-water Heaters or Economizers.—The modern type of fuel-economizer consists of a series of tubes, made up in sections, connected at the ends and placed in a brick chamber, through which the gases pass from the boiler to the chimney or fan. Feed-water is forced through the tubes, while the gases circulate around them. The difference between the economizers of different makes lies principally in the proportions, the design of end connections, and the position of the tubes. As is evident from the succeeding Table No. 74 and the explanation accompanying the same, no economizer is complete without some device for continuously or periodically removing the soot from the exterior of the tubes.

This type of economizer undoubtedly stands to-day as the most generally adopted means of abstracting the heat from the products of combustion. Its character, its cost, and to a

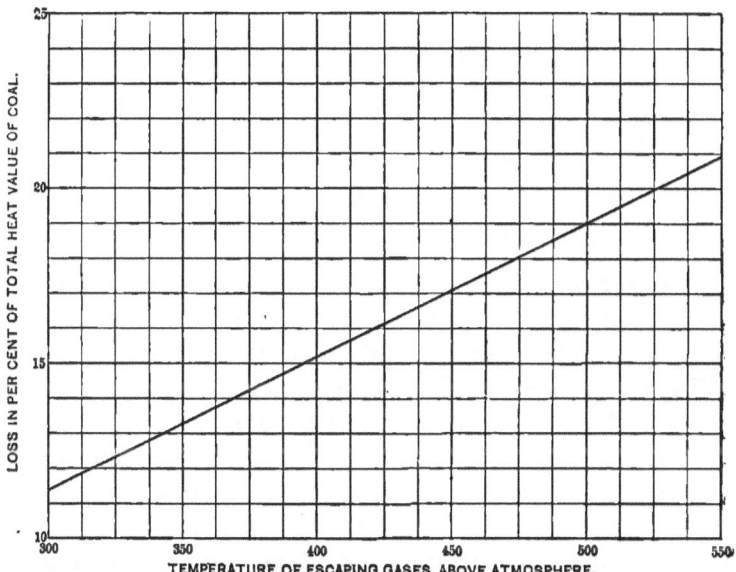

Fig. 10. LOSS OF EFFICIENCY DUE TO TEMPERATURE OF ESCAPING GASES.

certain extent its complication are serious detriments to its employment in connection with small boiler-plants.

Its earning power decreases with the cost of the fuel burned under the boilers in connection with which it operates. Assuming that approximately the same proportion of the total heat of the coal is utilized by the economizer with a cheap as with a high-priced coal, it is evident that its value then depends upon the number of heat-units represented by the amount of coal purchased for a given amount. The greater the value of a heat-unit the greater the cash saving by the economizer.

The economy resulting from the introduction of an economizer when the draft is sufficient naturally depends upon the normal temperature of the flue-gases escaping from the boilers, and of the feed-water supplied to the economizer. The percentage of gain may be determined by the following formula:

$$\text{Gain, in per cent,} = \frac{100\,(T-t)}{H-t}.$$

In which $T =$ heat-units in one pound of feed-water above 0° after heating;

$t =$ heat-units in one pound of feed-water above 0° before heating;

$H =$ heat-units in one pound of steam of boiler-pressure above 0°.

Table No. 71, calculated by this formula, indicates the saving under different conditions of feed-water, with steam of 70 pounds boiler-pressure. Of course the greatest economy will appear where the temperature of the flue-gases was originally the highest and that of the feed-water the lowest. Even with a low temperature of the flue-gases, an economizer will usually show results that warrant its introduction.

In Table No. 72 are presented the results of a number of tests by Barrus[*] of boiler-plants, of which an economizer in each case formed a part. In the first four cases, designated A, B, C, and D, the temperature of the gases as they leave the boiler is comparatively low,—averaging, 394°,—but the initial temperature of the feed-water is raised 92° on the average, and the evaporation increased 9.9 per cent per pound of coal. He shows that if this result is applied to a 1000 H.P. plant, burning Cumberland coal, the total annual saving in cost of coal would be about $2000.

---

[*] "Boiler-tests." George H. Barrus.

EFFICIENCY OF STEAM-BOILERS.

TABLE No. 71.

PERCENTAGE OF SAVING IN FUEL BY HEATING FEED-WATER. STEAM AT 70 POUNDS GAUGE PRESSURE.

Initial Temperature of Feed-water.	Temperature to which Feed-water is Heated.							
	100°.	110°.	120°.	130°.	140°.	150°.	160°.	170°.
35°	5.53	6.38	7.24	8.09	8.95	9.89	10.66	11.52
40	5.12	5.97	6.84	7.69	8.56	9.42	10.28	11.14
45	4.71	5.57	6.44	7.30	8.16	9.03	9.90	10.76
50	4.30	5.16	6.03	6.89	7.76	8.64	9.51	10.38
55	3.89	4.75	5.63	6.49	7.37	8.24	9.11	9.99
60	3.47	4.34	5.21	6.08	6.96	7.84	8.72	9.60
65	3.05	3.92	4.80	5.67	6.56	7.44	8.32	9.20
70	2.62	3.50	4.38	5.26	6.15	7.03	7.92	8.80
75	2.19	3.07	3.96	4.84	5.73	6.62	7.51	8.40
80	1.76	2.65	3.54	4.42	5.32	6.21	7.11	8.00
85	1.30	2.22	3.11	4.00	4.90	5.80	6.70	7.59
90	0.89	1.78	2.68	3.58	4.48	5.38	6.28	7.18
95	0.45	1.34	2.25	3.15	4.05	4.96	5.86	6.77
100	0.00	0.90	1.81	2.71	3.62	4.53	5.44	6.35

	180°.	190°.	200°.	210°.	220°.	250°.	300°.
35°	12.38	13.24	14.09	14.95	15.81	19.40	29.34
40	12.00	12.87	13.73	14.59	15.45	18.89	28.78
45	11.62	12.49	13.36	14.22	15.09	18.37	28.22
50	11.24	12.11	12.98	13.85	14.72	17.87	27.67
55	10.85	11.73	12.60	13.48	14.35	17.38	27.12
60	10.47	11.34	12.22	13.10	13.98	16.86	26.56
65	10.08	10.96	11.84	12.72	13.60	16.35	26.02
70	9.68	10.57	11.45	12.34	13.22	15.84	25.47
75	9.28	10.17	11.06	11.95	12.84	15.33	24.92
80	8.88	9.78	10.67	11.57	12.46	14.82	24.37
85	8.48	9.38	10.28	11.18	12.07	14.32	23.82
90	8.07	8.98	9.88	10.78	11.68	13.81	23.27
95	7.66	8.57	9.47	10.38	11.29	13.31	22.73
100	7.25	8.16	9.07	9.98	10.88	12.80	22.18

TABLE No. 72.

TESTS WITH ECONOMIZERS.

	Designation of Boiler.				
	A	B	C	D	E
Area of heating surface, boiler .. ..square feet	1894	4058	5592	3126	1880
Area of heating surface, economizer............square feet	1600	1920	1280	1600	1600
Temperature of gases leaving boiler...... ....degrees	376	361	403	435	618
Temperature of gases leaving economizer............degrees	231	254	299	279	365
Temperature of feed-water entering economizer.....degrees	95	79	111	84	88
Temperature of feed-water entering boiler ....degrees	175	145	169	196	225
Increased evaporation produced by economizer......per cent	10.5	7.0	9.3	12.8	29.0

An economizer equipment sufficient to secure the above result would probably cost from $7000 to $8000. The saving would represent an annual return of 24 to 28 per cent, certainly sufficient to warrant careful consideration.

Upon the same basis of calculation, the case designated E would show an annual return of about 75 per cent of the investment. Although this result is only approximate, it is sufficiently near the truth to indicate the indisputable economic advantage of the economizer. The heat-expenditure for producing the draft requisite to such results will be found to be less in the case of a fan than of a chimney, while the former will in addition possess advantages which make it much the more desirable of the two.

At the steam-plant of Cheney Brothers at South Manchester, Conn.,* which consists of 1000 horse-power of Babcock & Wilcox boilers provided with an economizer and mechanical-draft apparatus, 45 cubic feet of water is heated per minute and used in the boilers, while an additional 50 cubic feet is heated and utilized in the dye-house. The whole quantity is raised from an initial temperature of 112° to 211°. The heating of the feed-water alone is sufficient to cause a saving in the evaporative work of the boiler amounting to 10 per cent. So that the total saving, including the heat utilized in heating water for the dyehouse, is over 20 per cent. In a

---

* *The Engineering Record*, Jan. 6, 1894.

plant of 1000 horse-power running 10 hours per day, and using coal at $4.00 per ton, this represents an annual saving of about $3000 per year, an excellent return on the additional investment required to install the economizer plant.

As still further evidence of the saving in fuel which may be accomplished under working conditions by introducing economizers, the results of tests of nine plants by Roney[*] are presented in Table No. 73. They were all equipped with mechanical draft.

TABLE No. 73.

RESULTS OF TESTS OF MECHANICAL-DRAFT PLANTS AND ECONOMIZERS.

Plants Tested.	Temperatures.					Fuel Saving. Per cent.
	Gases Entering Economizer.	Gases Leaving Economizer.	Water Entering Economizer.	Water Leaving Economizer.	Gain in Temperature of Water.	
1	610°	340°	110°	287°	177°	16.7
2	505	212	84	276	192	17.1
3	550	205	185	305	120	11.7
4	522	320	155	300	145	13.8
5	505	320	190	300	110	10.7
6	465	250	180	295	115	11.2
7	490	290	165	280	115	11.0
8	495	190	155	320	165	15.5
9	595	299	130	311	181	16.8

The influence of soot, and the necessity of frequent if not continuous cleaning of the surfaces of an economizer, was clearly shown by Grosseteste[†] in a three weeks' test with smoky coal upon a Green economizer, consisting of a series of vertical pipes arranged to be cleaned externally by automatic scrapers. The apparatus had been at work for seven weeks continuously without having been cleaned, and had accumulated a half-inch coating of soot and ash. It was

[*] "Mechanical Draft." W. R. Roney. Transactions American Society of Mechanical Engineers, vol. xv.
[†] Bulletin de la Société Industrielle de Mulhouse, vol. xxxix.

observed in this condition throughout the first week. During the second week it was cleaned twice every day, but during the third week, after having been cleaned on Monday morning, it was worked continuously without further cleaning. The results presented in Table No. 74 show the necessity of cleaning.

TABLE No. 74.

INFLUENCE OF THE STATE OF THE SURFACE OF AN ECONOMIZER.

Time. (February and March.)	Temperature of Feed-water.			Temperature of Gaseous Products.			Coal Consumed per Hour.	Water Evaporated from 32° per Hour.	Water per lb. of Coal.
	Entering Economizer.	Leaving Economizer.	Difference.	Entering Economizer.	Leaving Economizer.	Difference.	lbs.	lbs.	lbs.
First week *	73.5°	161.5°	88.0°	849°	261°	588°	214	1424	6.65
Second week	77.0	230.0	153.0	882	297	585	216	1525	7.06
Third week: Monday..	73.4	196.0	122.6	831	284	547	} 213	1428	6.70
Tuesday..	73.4	181.4	108.0	871	309	562			
Wednesd.	79.0	178.0	99.0	......	......	......			
Thursday	80.6	170.6	90.0	952	329	623			
Friday...	80.6	169.0	88.4	889	338	551			
Saturday.	79.0	172.4	93.4	901	351	550			

* The averages for the first and second weeks are exclusive of Monday.

Air-heaters or Abstractors.—The air-heater, heat-abstractor, warm-blast apparatus, or hot-draft apparatus, as it is variously called, generally consists of some arrangement of pipes, through or across which the hot gases pass directly from the uptake, heating thereby the air-supply for the furnace, which passes respectively across or through the pipes; the object being to abstract from the gases as much heat as is practicable and transfer it to the air before it enters the furnace, thereby securing a higher temperature and increased evaporative efficiency.

Such an apparatus, is virtually the equivalent in results of an economizer, and is the only practical means of reducing the waste of heat in the flue-gases when large quantities of warm water are not in demand, as must be the case if an economizer is to show efficient results. The ideal arrangement, however, consists of a combination of economizer and abstractor, whereby the air-supply to the furnace may be heated as well as the feed-water for the boiler, and all the heat practicable thus abstracted from the flue-gases. For such results chimney-draft is ordinarily inadequate and mechanical means must be resorted to to overcome the increased resistance.

Doubtless the most comprehensive test ever conducted upon an apparatus of this character was that undertaken by Hoadley,* in the interest of a number of mill-owners, and intended to determine the efficiency of the Marland apparatus for heating the air-supply. The original apparatus was applied to a single horizontal return tubular boiler, 60 inches in diameter, with 65 tubes $3\frac{1}{2}$ inches in outside diameter, and 20 feet long. This boiler was of the regular type in use in the Pacific Mills, Lawrence, Mass., where the tests were made, except that upon its top were placed two abstractors, one upon either side, 3 feet apart, extending the entire length of the boiler. Each abstractor contained 120 lap-welded tubes, 2 inches in outside diameter and 20 feet long, inclosed in a brick setting. Surrounding each pipe was a 3-inch tube of thin iron. By proper arrangement of the heads into which the 2-inch pipes were expanded, in connection with the uptake, a passage was provided for the flue-gases through these pipes and thence to the blower which produced the requisite draft. The 3-inch tubes were shorter than the 2-inch, and were arranged at their ends so that air for the furnace could pass to them and thence through the annular space between the two tubes, becoming heated by the gases in the inner tube. This apparatus was known as Warm-blast No. 1. Alongside this boiler, and operating under the same conditions, was a boiler of the regular type, designated in the report as Pacific Boiler.

---

* "Warm-blast Steam-boiler Furnace." J. C. Hoadley.

170   STEAM-BOILER PRACTICE.

Extended experiment having shown Warm-blast No. 1 to be incapable of reducing the temperature of the escaping gases below 160°, the Pacific Boiler was, accordingly, converted into Warm-blast No. 2 by placing upon its top two abstractors differing from the previous ones, and constructed substantially as follows: 2-inch spiral-locked tubes were provided for the passage of the hot gases, while the 3-inch tubes were replaced by a series of deflectors set at right angles to the tubes, which passed through them. These deflectors were so arranged that air entering at the top must descend across and among the 2-inch tubes, which had 1-inch spaces between them, pass under the first deflector, then rise in the same manner and pass over the second deflector, and so on, until the air passed to the ash-pit.

The comparative temperatures found to exist in connection with the Pacific and Warm-blast No. 1 boilers, properly reduced for comparison, are presented in Table No. 75.

TABLE No. 75.

COMPARATIVE TEMPERATURES IN PACIFIC BOILER AND WARM-BLAST BOILER NO. I.

Location at which Temperature was Taken.	Temperatures.		
	Pacific Boiler.	Warm-blast Boiler.	Difference.
In heart of fire............................	2493°	2793°	300
At bridge-wall.............................	1340	1600	260
At pier...................................	895	1050	155
In smoke-box..............................	373	375	2
Air admitted to furnace....................	32	332	300
Steam and water in boiler..................	300	300	0
Gases escaping to chimney..................	373	162	211
External air...............................	32	32	0
Gases cooled, warm-blast boiler............			213
Air warmed, warm-blast boiler..............			300

These figures alone seem to point to the efficiency of the warm-blast apparatus; but Table No. 76, comprising the important economic results, serves to indicate more definitely the relative efficiency of the various arrangements, and to prove the marked advantage of the warm-blast arrangement. Careful tests showed that the power consumed in driving the blower was about 1 per cent of the whole power produced by the boiler in combination with a good steam-engine. This should be compared with the much larger expenditure required to produce the draft by means of a chimney.

TABLE No. 76.

RESULTS OF TESTS WITH PACIFIC BOILERS AND WARM-BLAST BOILERS.

Items.	With Anthracite Coal.			With Bituminous Coal.	
	Pacific Boiler.	Warm-blast No. 1.	Warm-blast No. 2.	Pacific Boiler.	Warm-blast No. 1.
Mean temperarure of external air, days.....degrees	78.3	34	49	71	34.2
Temperature of smoke-box.. ....degrees	368.3	396.9	377	376.9	397.4
Temperature of escaping gases. ....degrees	368.3	189	164	376.9	196
Gases cooled by abstractors...............degrees	0	207.9	213	0	201.4
Air warmed by abstractors....... ....degrees	0	303.7	285	0	315.5
Temperature of air supplied to furnace.....degrees	78.3	337.7	334	71	349.5
Temperature of steam......................degrees	297.5	361.1	291.2	297.3	322.6
Loss at chimney.......... .... per cent	17.75	15.00	12.83	17.03	14.24
Loss by radiation from brickwork.........per cent	2.64	4.00	4.00	3.39	4.00
Loss by imperfect combustion.............per cent	2.13	0.63	1.43	2.85	1.06
Total loss by above three causes........ ...per cent	22.52	19.63	18.26	23.27	19.30
Pounds of flue-gases per pound of coal.............	22.39	23.49	24.17	25.23	28.37
Efficiency, reduced to common basis.......per cent	68.87	78.18	81.43	64.61	77.59
Difference of efficiency, points gained by warm-blast over Pacific boiler............................	....	9.31	12.56	....	12.98
Ratio of gain to the larger quantity $\left(\frac{9.31}{78.18} = 11.9\%\right)$	....	11.9	15.4	....	16.7
Ratio of gain to the smaller quantity $\left(\frac{9.31}{68.87} = 13.5\%\right)$	....	13.5	18.2	....	20.1

The results obtained with the Ellis and Eaves system of preheating the air in connection with boilers at the American Line Pier, New York, N. Y.,* are presented in Table No. 77. These are particularly interesting as showing the temperature and the intensity of draft at different points. The boilers were of the double-ended Scotch type, one fitted with plain tubes and the other with Serve patent ribbed-steel tubes; all tubes were provided with retarders. The waste gases were carried

---

* *The Engineering Record*, Jan. 5, 1895.

from the breeching back over the boilers through two air-heating chambers, and finally passed from these through the necessary connections to an 8-foot fan discharging into a 70-foot stack. The air-heating chambers contained a number of 3-inch tubes, each about 12 feet long, through which the air that supplied the furnaces was drawn. The air was highly heated in its passage by the waste gases from the boiler, and delivered at a high temperature both above and below the fires. The coal used was No. 1 buckwheat Susquehanna.

TABLE No. 77.

RESULTS OF TESTS OF ELLIS & EAVES INDUCED-DRAFT SYSTEM AT AMERICAN LINE PIER 14, N. R., NEW YORK, N. Y.

Items.	Number and Conditions of Test.		
	No. 1. No. 1 Boiler, Plain Tubes with Retarders.	No. 2. No. 1 Boiler, Plain Tubes with Retarders.	No. 3. No. 2 Boiler, Serve Tubes with Retarders.
Date.... ..1894	Sept. 28	Oct. 5	Sept. 29
Duration of test.... ..hours	6	6	6
Average steam-gauge pressure............. pounds	96	95	96
Average temperature of feed-water ... ... deg. F.	157	166	165
Average revolutions of fan per minute	507	472	468
Total water evaporated .. ..lbs.	72661	59708	71697
Total coal burned... ... .lbs.	11000	8920	8715
Total combustible consumed ....lbs.	9746	7780	7717
Coal burned per square foot of grate per hour,.lbs.	52.08	42.23	41.26
Water evaporated per pound of coal from and at 212°... ... ...lbs.	7.228	7.274	8.939
Water evaporated per pound of combustible from and at 212°......... .. .. :. ..lbs.	8.165	8.332	10.097
Average temperature at fan outlet..........deg. F.	625	527	456
Average temperature at fan inlet ......'....deg. F.	650	547	472
Average temperature air down-take........deg. F.	400	347	315
Average vacuum as fan inlet... ...inches of water	7.33	6.61	6.45
Average vacuum over fires ....inches of water	3.88	3.46	3.56
Average vacuum under fires ......inches of water	1.72	1.38	1.48
Average vacuum at air down-take..inches of water	0.62	0.58	0.50

Increased Tube-heating Efficiency.—With fire-tubes of a given length the amount of heat transmitted to the water within the boiler must be dependent upon the temperature and velocity of the gases, the amount of surface exposed, and the completeness with which they are forced into contact with it. In other words, with the same velocity and temperature, a given length of tube will be efficient in proportion as it presents absorbing surface for receiving the heat of the gases, and as those gases are compelled to come in contact with it. Two methods are in use for accomplishing this result. The first consists in fitting within a regular boiler-tube a strip of thin sheet-iron, equal in width to the internal diameter of the tube, and twisted so as to form a helix of long pitch, making only two or three turns in the length of the tube. The effect of this arrangement—the strip being known as a "retarder" —is to break up the current of gas and cause all portions of its volume to touch the inner surface of the tube. At the same time the retarder itself is intensely heated, and rapidly radiates its heat through the tube to the water. The dual effect of the retarder is to materially increase the evaporative power and the efficiency of the boiler.

The economic result of the use of retarders is shown in Table No. 78 by the tests of Whitham * upon a 100 H.P.

TABLE No. 78.

REDUCTION IN TEMPERATURE OF FLUE-GASES AND IN COAL-CONSUMPTION BY THE USE OF RETARDERS.

Horse-power Developed.	Reduction in Temperature of Flue-gases. Degrees Fahr.	Reduction in Coal-consumption. Per cent.
52	20	0.0
75	53	0.0
100	32	3.2
125	46	4.0
150	19	3.3
170	59	3.6
200	36	4.1
225	26	8.6
239	123	18.4

horizontal tubular boiler operated at from about 50 per cent below to about 140 per cent above its rated capacity. The evident result is a reduction in the temperature of the flue-

---

* "The Effect of Retarders in Fire-tubes of Steam-boilers." J. M. Whitham. Transactions American Society of Mechanical Engineers, vol. XVII.

gases, with a corresponding decrease in the coal-consumption. But, obviously, as the name "retarder" implies, this result cannot be attained without an increase in the draft. In this connection Whitham presents results, given in Table No. 79, showing the different drafts and resistances.

TABLE No. 79.

DRAFT AND RESISTANCE WHEN RETARDERS ARE USED.

Items.	Draft or Resistance in Inches of Water.
Furnace-draft...	0.30
Resistance of pass under boilers and through tubes without retarders..	0.27
Total draft of stack if no top pass is used.................	0.57
Resistance due to having retarders........................	0.31
Total draft if there is no return pass and retarders are used.	0.88
Increased resistance due to return pass over top of boilers..	0.07

Whitham's general deductions regarding retarders are that they interpose a resistance varying with the rate of combustion; that they reduce the temperature of the flue-gases, and increase the effectiveness of the heating-surface; that they should not be used where the draft is small; that they can be used to advantage in plants using a fan, and that they may show from 5 to 10 per cent advantage whenever the boiler-plant is pushed and the draft is strong.

The second method of increasing the efficiency of fire-tubes is more direct, and consists in a special construction of the tube itself. Such is the case in the Serve tube, which is outwardly cylindrical, but from whose inner or fire surface a number of equidistant radial ribs parallel to the axis converge toward the centre. The radial length of the ribs is usually about one fifth of the external diameter of the tube, and they are seven or eight in number, according to the external diameter. The superior economy of these tubes is accounted for upon the theory that the ribs break up the column of gases, and by means of their extended surface extract heat from all parts of it. A six-day comparative test of the efficiency of plain and ribbed tubes, under practically identical conditions, was made by Roelker,* and the general results are presented in Table No. 80. In some cases retarders are used with these tubes, securing thereby even better contact of the air, because of its being forced to pass through the spaces between the retarder and the ribs. Comparative results are presented in Table No. 126

TABLE No. 80.

COMPARATIVE TESTS OF EFFICIENCY OF RIBBED AND PLAIN FIRE-TUBES.

Duration of Tests.	Manner of Producing Draft.	Intensity of Draft. Inches Water.	Pounds of Water Evaporated per Pound of Coal.	
			Plain Tubes.	Ribbed Tubes.
Two 8-hour tests...............	Natural	1/8	5.08	7.35
Two 8-hour tests...............	Mechanical	1/2	5.98	7.60
Two 8-hour tests...............	Mechanical	7/8	4.68	6.75
One 8-hour test................	Mechanical	13/16	....	7.41
One 8-hour test................	Mechanical	19/16	....	6.52

Of course the highest relative efficiency of such devices will be shown when the tubes are comparatively short and the gases with plain tubes are rejected at a high temperature. But under all ordinary conditions there can be no question of their efficiency. They are of especial advantage when space or the design of the boiler forbids the convenient introduction of additional heating-surface in any other manner, as is usually the case in marine boiler-practice. They cannot, however, be advantageously used when the draft is small, as has already been evidenced.

Another device designed to utilize the waste heat of the gases for warming the air supply is the Eaves helical boiler.

---

* "Serve's Ribbed Boiler-tube." Passed Assistant Engineer G. S. Willits, U. S. Navy. Journal of American Society of Naval Engineers, August, 1891.

This is of the marine type with a double inclosing case of sheet steel. The gases escape into the annular space between the boiler-shell and the inner envelope, and are caused to pass around the boiler in a helical direction, thus giving them time to part with a portion of their heat to the air as it passes in a similar manner between the inner and outer envelopes. Thence it enters the furnaces at a high temperature, while the gases are drawn through the draft-fan in decreased volume and temperature. The results of tests of this device applied to a Scotch boiler 10.5 feet diameter by 10.5 feet long are presented in Table No. 81.

TABLE No. 81.

RESULTS OF TEST OF EAVES' HELICAL-DRAFT BOILER.

Items.	First Trial.	Second Trial.
Duration ....hours	7	7
Total coal burned..........................pounds	7581	8022
Total water evaporated ....pounds	71000	71500
Temperature of feed-water...........degrees Fahr.	54	50
Steam-pressure............pounds per square inch	43.4	45
Revolutions of fan-engine per minute...............	508	520
Temperature of air at side valve.......degrees Fahr.	234	259.5
Temperature of gases at inlet of fan....degrees Fahr.	309	353.8
Temperature of gases in smoke-box................	Melted bismuth, not lead.	Melted bismuth, not lead.
Vacuum under grate bars...............inches of water	0.75	0.64
Vacuum over fires.....................inches of water	0.82	0.81
Vacuum at base of chimney............inches of water	4.59	4.59
Vacuum above fan outlet.............inches of water	0.38	0.38
Velocity of air per minute under grate bars........feet	1476	1362
Velocity of air per minute through outer casing....feet	257	227
Temperature of the air entering outer casing, degrees Fahr..	78	67
Coal burned per square foot of grate per hour...pounds	33.84	35.81
Water evaporated per pound of coal ....pounds	9.36	8.91
Water evaporated per lb. of coal from and at 212°..pounds	11.12	10.63
Calorific value of coal used, expressed in pounds of water evaporated from and at 212°....................	13.6	13.6
Efficiency of boiler.........................per cent	82	78.3

Mechanical Stokers.—The higher efficiency attained when the firing is in small amounts at frequent intervals, and when the fire is carefully maintained in the best possible condition, points to the results which should ensue from the employment of a proper method of continuously feeding the fire by mechanical means. The mechanical stoker, as a substitute for hand-firing, possesses many advantages, but they can only be realized when the stoker is properly suited to its particular work and is intelligently operated. Its advantages may be summarized as—

First. Adaptability to the economical combustion of the cheapest grades of fuel.

Second. Saving in labor of firing.

Third. Economy in combustion even with forced firing, under proper management.

Fourth. Constancy and uniformity of the furnace conditions.

Fifth. Smokelessness.

Three principal types of mechanical stokers are to be found in use.

The under-feed, by which the fuel is, by screw or plunger, forced upward from beneath. In the common forms the fuel is thus supplied along the centre of the length of the grate, and as it is forced upward falls over to the sides and thus forms a long mound, thin at the sides of the grate and of considerable thickness in the centre. This thickness necessitates a very strong under-grate blast, which can only be secured by the use of a blower, and is generally applied in greatest volume at or near the centre.

The inclined over-feed type generally consists of a sloping grate, the highest portion being at the front of the boiler, where the coal is fed. The grate-bars are usually so constructed and arranged that they may be periodically moved so as to feed the fuel along and down the surface. The motive power by which this practically continuous movement and feeding process is maintained is, in most cases, derived from a small independent engine, although hand-power is sometimes used in small plants. Evidently this latter method of operating results in a much less frequent movement of the coal. Excellent results with this form of mechanical stoker can be obtained when forced draft is used and the air is admitted either through the grate-bars themselves, or between them from the space beneath. With either arrangement there is an excellent opportunity for the most perfect distribution of the air and its intimate contact with the fuel.

The third type of mechanical stoker consists of a chain-grate upon which the fuel is fed at the front of the boiler, and which by its slow progress toward the bridge-wall gives the fuel an excellent and undisturbed opportunity for complete combustion. It is particularly adapted for the lowest grades of fuel, and in its most perfect form is so arranged that the fuel, at different points in its progress, receives its air-supply in different amounts and under different pressures, each best suited to the given stage of the combustion. The proper introduction and regulation of the air requires that it should be supplied by a blower, which thus forms an inherent part of such a plant. Comparisons between mechanical stoking and hand-firing, as well as between different forms of mechanical stokers, demand that the conditions shall be practically identical. In Table No. 82 are presented the results of

TABLE No. 82.

TESTS OF MECHANICAL STOKERS.

	A		B		C		D	
	Hand.	Stoker.	Hand.	Stoker.	Hand.	Stoker.	Hand.	Stoker.
Coal per hour, pounds....	2022	1422	2022	1800	428	432	857	771
Coal per hour per sq. ft. grate, pounds...........	25.3	28.4	25.3	29.0	....	....	29.1	30.9
Water evaporated per hour per pound of coal from and at 212°, lbs...	7.11	8.70	7.11	9.12	7.68	8.80	8.81	10.29
Water evaporated per hour per pound of combustible from and at 212°, pounds..	7.77	9.67	7.77	10.07	8.96	10.01	....	....

several comparative tests made under such conditions, each with a different form of mechanical stoker. They serve to show the undeniable economy resulting from this method of feeding coal. Although several forms of mechanical stokers are included in the list, the tests are, for obvious reasons, designated only by distinguishing letters, each letter covering the results with both hand-firing and mechanical stoking under similar conditions.

Constant attention is necessary in mechanical firing in order to regulate the rate of feed to the rate of evaporation; but the total amount of labor is far less than that required in hand-firing. When bituminous coal is mechanically fired there can be but little question that the plant, if of reasonable size, will operate with sufficient economy to pay a good return on the extra investment required for the stoking apparatus.

Powdered-fuel Furnaces.—Coal, in the form of dust, fed to the boiler-furnace in a current of air, has to some extent been employed for the purposes of steam-generation. The arrangement usually comprises a device for reducing the coal to an impalpable powder. It is then fed, together with air ordinarily supplied by a fan, into the front of the furnace. Theoretically this method appears to have certain advantages which should make it successful. There is opportunity for instantaneous combustion, and the most intimate contact of the air, whereby the minimum amount may be employed. There should be no loss by decrepitation; but this is more than offset by the tendency to blow the dust in an unconsumed state directly up the chimney. The results indicate, however, that with such methods as have been tested the gain, if any, is more than counterbalanced by the added expense.

CHAPTER VII.

RATE OF COMBUSTION.

Rate of Combustion.—The rate at which fuel is consumed in steam-boiler practice is usually expressed in pounds per hour per square foot of grate-surface. Evidently this rate must vary greatly according to the conditions governing in a given case. Although the draft is by far the most important factor, yet the amount of coal burned per unit of area also depends, to a certain extent, upon the total area and arrangement of the grate-surface, the method of firing, the kind of fuel, and other factors of less importance.

In the early days of the steam-boiler the rate of combustion, like the speed of the first steam-engines, was extremely low. The grates were large, the draft comparatively light, and forcing of the boilers less common than at the present day. The general rates of combustion, as given by Rankine[*] for various types of boilers and conditions of draft, are presented in Table No. 83.

Since these figures were first given the rates have increased, so that 15 to 20 pounds is not at all uncommon in factory boilers, under exceptionally strong draft; while 25 to 40 pounds is the usual marine practice, and 60 to 125 pounds are burned in the boilers of torpedo-boats. In fact, modern steam-engineering practice is constantly looking toward higher rates of combustion as an accompaniment to higher steam-pressures and engine-speeds, in the attempt to attain increased efficiency.

---

[*] "A Manual of the Steam-engine." W. J. M. Rankine.

TABLE No. 83.
RATES OF COMBUSTION.

Conditions.	Method of Producing Draft.	Pounds of Coal per Square Foot of Grate per Hour.
Slowest rate of combustion in Cornish boilers...	Chimney	4
Ordinary rates in Cornish boilers................	"	10
Ordinary rates in factory boilers................	"	12 to 16
Ordinary rates in marine boilers.................	"	16 to 24
Quickest rates of complete combustion of dry coal, the supply of air coming through the grate only..	"	20 to 23
Quickest rates of complete combustion of caking coal, with air-holes above the fuel to the extent of 1/36 area of grate.....................	"	24 to 27
Locomotives...................................	Blast-pipe or fan	40 to 120

The application of forced draˑt to a furnace affords a means of obtaining a higher rate of combustion of fuel per square foot of grate-surface per hour than is conveniently obtainable with natural draft. The rate of combustion obtained in practice varies, with the intensity of the draft, from 30 to 200 pounds of coal per square foot of fire-grate surface per hour. A moderate rate of forced combustion is from 35 to 50 pounds of coal per square foot of grate per hour.

Relation of Grate-surface to Heating-surface. — The ratio existing between the areas of grate- and heating-surface in various types of boilers has already been presented in Table No. 69. The influence of this ratio on the evaporation has also been indicated. As the economy of the boiler is usually expressed in the number of pounds of water evaporated per pound of coal, it is evident that, the surface-ratio and the rate of combustion also being known, the relative capacity of the boiler can be determined and expressed in pounds of water evaporated per square foot of heating-surface. In fact, this should be the ultimate basis of comparison rather than the rate of combustion; for the influence of the latter is dependent upon the surface-ratio, which may vary considerably even in boilers of the same type. The results of this influence are shown in Table No. 84, in which, with a constant fuel-efficiency of 10 pounds of water evaporated from and at 212° per pound of coal, the evaporation per square foot of heating-surface has been calculated for different rates of combustion under different surface-ratios. Of course this table is purely theoretical, and applies only when the efficiency is 10 pounds, but it indicates relative values.

TABLE No. 84.

RATES OF EVAPORATION PER SQUARE FOOT OF HEATING-SURFACE.

Surface-ratio.	Pounds of Coal per Hour per Square Foot of Grate.								
	5	10	15	20	25	30	35	40	45
1 : 25	2.00	4.00	6.00	8.00	10.00	12.00	14.00	16.00	18.00
1 : 30	1.67	3.33	5.00	6.66	8.33	10.00	11.67	13.33	15.00
1 : 35	1.43	2.86	4.29	5.72	7.15	8.58	10.00	11.44	12.87
1 : 40	1.25	2.50	3.75	5.00	6.25	7.50	8.75	10.00	11.25
1 : 45	1.11	2.22	3.33	4.44	5.55	6.66	7.77	8.88	10.00
1 : 50	1.00	2.00	3.00	4.00	5.00	6.00	7.00	8.00	9.00
1 : 55	0.91	1.82	2.73	3.64	4.55	5.46	6.36	7.27	8.17
1 : 60	0.83	1.67	2.50	3.33	4.17	5.00	5.63	6.67	7.50

It is evident that, with a given boiler, it is a comparatively simple matter to so change the setting as to permit of the use of grates of a different area, thus altering the surface-ratio; and, further, that a reduction of grate-surface with the maintenance of the same rate of combustion is relatively equivalent to an increase in the heating-surface without change of the original grate. A reduction of grate-area, without corresponding increase in the rate of combustion, such that the total amount of fuel consumed remains the same, must of necessity reduce the rate of evaporation per square foot of heating-surface, and consequently the capacity of the boiler. Under such conditions the evaporative efficiency may be improved, but the boiler be actually incapable of performing its previous amount of work. It is by such a combination of circumstances that the unprincipled advocates of certain so-called coal-saving devices have sometimes been able to show an increased evaporation per pound of fuel, which, while satisfactory in itself, was not secured under the ordinary working conditions, and might possibly have been as easily secured without the device. The capacity may be maintained, even with reduced grate-area, by correspondingly increasing the rate of combustion. The limit to such arrangements is naturally set by the amount of fuel which can be fired and maintained in good condition per square foot of grate; while the evaporative capacity of the heating-surface is limited by the ability of the steam-bubbles to readily escape from its surface, which ability in turn is largely dependent upon the arrangement of the surface.

Economy of High Rates of Combustion.—It has been shown in the preceding chapter that with a *constant* surface-ratio the evaporative efficiency of the boiler decreases as the rate of combustion is increased. But an increase of the rate, while the surface-ratio remains the same, is in effect an increase in the total quantity of coal consumed. The more the coal consumed the more the air required; hence the logical deduction that the efficiency will be reduced. For the larger volume of gases travelling at higher velocity will impart relatively less heat to the exposed surfaces and enter the chimney at a higher temperature. But in practice certain factors sometimes affect these theoretical considerations, so that the rate of combustion upon a given grate may be increased without appreciable adverse economic effect.

It is to be noted that when the surface-ratio is constant the rate of combustion becomes practically a direct measure of the total amount of fuel consumed. If, however, the ratio be changed, as for instance by reducing the grate-area, the total consumption can only be maintained by increasing the rate of combustion per square foot. Under this condition it is certainly evident that at least no greater amount of air will be required per pound of coal. In fact, experience shows that ordinarily the amount of air required will actually be reduced. This naturally results in an increase in the efficiency. Upon this fact rests one of the important advantages claimed for mechanical draft, for by its means may be produced the intensity of draft requisite to high combustion-rates.

The conditions under which this increased efficiency can be secured must, however, be clearly understood. An increase in the rate of combustion, when it is accompanied by a greater coal-consumption, as would be the case where the surface-ratio remains constant, is by no means always conducive to economy. But when the total amount of coal consumed remains the same, and the increased combustion-rate is secured by a reduction of grate-area and corresponding increase in the surface-ratio, a higher efficiency is the natural result. This is shown by Table No. 70, the figures in which give relative values.

As there indicated, the efficiency of the fuel, or the water evaporated from and at 212° per pound of coal, is 10.23 pounds for a stationary boiler, where the rate of combustion is 30 pounds per square foot of grate and the surface-ratio is 30. This is equivalent to burning $30 \div 30 = 1$ pound of coal per square foot of heating-surface per hour. If the surface-ratio were 50, and the coal-consumption per square foot of heating-surface remained the same,—namely, 1 pound,—the rate per square foot of grate would be $50 \times 1 = 50$ pounds. The table shows that with these conditions of surface-ratio and rate of combustion the evaporation would be 10.67 pounds, an increase of about 4 per cent. The high rates are chosen for illustration only because they avoid interpolation in the table; but the same principle holds throughout. For instance, a rate of 25 pounds per square foot of grate, with a ratio of 50, gives about 9 per cent higher efficiency than a rate of 15 pounds and a ratio of 30, although the coal-consumption per square foot of heating-surface is the same.

The principal if not the sole cause for this increase in

efficiency is to be found in the decreased supply of air which is required per pound of coal when the rate of combustion per square foot of grate is raised. The reason of the decreased requirement appears evident in the fact that the higher rate of combustion necessitates a deeper fire, and that the air supplied is, therefore, compelled to come in contact with a greater amount of fuel, and afforded a better opportunity to promote perfect combustion. The intensity of the fire is increased, its temperature is higher, more heat is radiated to the exposed boiler-surfaces, and more is taken up by the gases. Furthermore, the diminished superficial area of the grate and of the exposed interstices between the fuel necessitates a higher velocity to secure the admission of a given volume of air. This increased velocity in turn requires greater draft or air-pressure. The volume at a given temperature passing through the coal is proportional to the velocity, but the pressure varies as the square of the velocity. Therefore, if a given grate be reduced one half, and the rate of combustion doubled, so as to maintain the same total consumption, the same volume of air would have to travel through the exposed interstices at twice the velocity. But the pressure or vacuum would be four times as great, and, as a consequence, the air would be forced or drawn into spaces between the fuel which it could not reach under lesser impelling force. Much more intimate contact and distribution are the results. Less free oxygen passes through the fuel-bed unconsumed, and for a given supply of air a higher efficiency of the fuel is attained. But the high pressures necessary to such results seldom exist or are attainable where the ordinary chimney is depended upon for the production of the draft. In fact, the present rates of combustion common in factory practice are such largely because of the inability of a chimney of moderate height and cost to economically provide draft of sufficient intensity for higher rates. For this reason forced or mechanical draft has always been considered in a sense separate and apart from chimney-draft, and until recently has been regarded principally in the capacity of producing draft-pressures beyond the limits of the ordinary chimney.

Nearly forty years ago Rankine wrote that "in furnaces where the draft is produced by means of a blast-pipe, like those of locomotive-engines, or by means of a fan, the quantity of air required for dilution, although it has not yet been exactly ascertained, is certainly much less than that which is required in furnaces with chimney-drafts; and there is reason to believe that on an average it may be estimated at about *one half* of the air required for combustion."

Hutton* bases his estimate of increased efficiency with mechanical draft upon the decreased amount of air required therewith, and the resulting higher furnace-temperature. Accepting 24 pounds of air as necessary per pound of coal under natural draft, he shows the furnace-temperature to be 2926° with perfect combustion, giving an efficiency of 66 per cent. Accepting 18 pounds as required under forced draft, the furnace-temperature is shown to be 3686°, and the efficiency 76 per cent.

This subject has not been so carefully investigated as its importance warrants, but it is reasonably well established that, under ordinary conditions, a decreased amount of air is required with rapid rates of combustion. Owing to the influence of surface-ratios, kind of fuel, construction of the furnace, and the like, any absolute basis of comparison is, however, practically impossible.

Recent extended experience with American boilers and coals, as shown by the following quotations, points clearly to the substantial maintenance under proper operation of the evaporative efficiency with widely varying rates of combustion. Low, in the report of the Committee on Data of the National Electric Light Association,† states: "I have plotted the water evaporated per hour per square foot of heating-

---

* "Steam-boiler Construction." Walter S. Hutton.
† *The Engineering Record*, June 26, 1897.

surface, and the pounds of water evaporated from and at 212° per pound of combustible, as determined by thirty different tests on Babcock & Wilcox boilers, and proved that practically as good results are obtained at over 5 pounds per square foot of heating-surface as at 1.75 pounds, and the intermediate tests show no dependence on the rate of evaporation. I have plotted in the same way all of the tests of which I could find a record, and in this wide range still no evidence is apparent of any dependence of the boilers represented on the rate of evaporation within the range covered. This means that the variation of efficiency of boilers within the range here comprised is less than the variations due to different firing, etc.

"Bryan, in a paper read before the Engineers' Club of St. Louis, gives the data and tests in which a battery of horizontal tubular boilers were forced to more than double their rating with only the following improvement of their efficiency: [In brief the tabulated results show the coal per square foot of grate-surface to range between 18.074 and 43.68, that per square foot of heating-surface between 0.332 and 0.803, the water evaporated per square foot of heating-surface between 2.43 and 5.235, and the evaporation per pound of combustible from and at 212° between 9.27 and 8.827. In percentage of rated capacity the range is between 100.2 and 219.83, while the efficiency percentage of heat utilized varies between 76.38 in the former and 68.83 in the latter.] Here is a battery of boilers which were forced to nearly double their rated capacity with a decrease of only 6 per cent in their efficiency, and which could doubtless have been diminished to one third of their rated capacity with a less impairment still. In other words, with good management and an adaptation to conditions, these boilers would have taken care of a maximum load six times the minimum without suffering extremely."

Clark[*] states that "the proportion of surplus air required appears to diminish as the rate of combustion and the general

---

[*] "The Steam-engine." D. K. Clark.

temperature in the furnace are increased," and presents the results here given in Table No. 85 as evidence of the truth of

TABLE NO. 85.

SURPLUS AIR WITH DIFFERENT RATES OF COMBUSTION.

Kind of Boiler.	Coal Consumed per Square Foot of Grate per Hour. Pounds.	Surplus Air Per cent.
Cornish................................	2 to 4	100
Delabèche and Playfair............	10 to 16	25 to 50
Longridge.........................	20 and upwards	9¾

this statement. Although the surplus air with the Longridge boiler appears extremely low, yet the experiments of Whitham, given in Table No. 86, although conducted under special conditions, tend to confirm its probability.

TABLE NO. 86.

AIR-SUPPLY WITH DIFFERENT RATES OF COMBUSTION ON A WILKINSON AUTOMATIC MECHANICAL STOKER.

Buckwheat Coal Burned per Hour per Square Foot of Stoker-grate. Pounds.	Air Theoretically required to Burn One Pound of Buckwheat Coal. Cubic Feet.	Air Actually Supplied to Burn One Pound of Buckwheat Coal. Cubic Feet.	Percentage of Excess or Deficiency of Air Supplied.
12.0	125	232	+ 85.6
18.0	125	157	+ 25.6
25.2	125	132	+ 5.6
32.5	125	123	− 1.6
41.5	125	111	− 11.2
45.4	125	111	− 11.2

Although it is commonly accepted that the air-supply with chimney-draft is about 300 cubic feet (approximately 24 pounds per pound of coal where the combustion-rate is from 10 to 15), yet there is in practice a wide variation from this standard. Thus the tests of Messrs. Donkin and Kennedy, already presented in Table No. 13, show a range between

16.1 pounds, and 40.7 pounds per pound of coal; that is, an excess of 56 to 328 per cent. In marine practice, with forced draft, a combustion-rate of 30 to 40 pounds may be easily maintained with a supply of 225 cubic feet of air per pound of coal.

Greater care in the distribution of the air and in maintaining the condition of the fire is necessary to a successful reduction in the air-supply. This is very clearly evidenced by the remarkable results of tests by Whitham* upon automatic mechanical stokers. Those relating to the Wilkinson stoker, under different rates of combustion, are presented in Table No. 86 and also in Fig. 11. It is to be noted that there is an

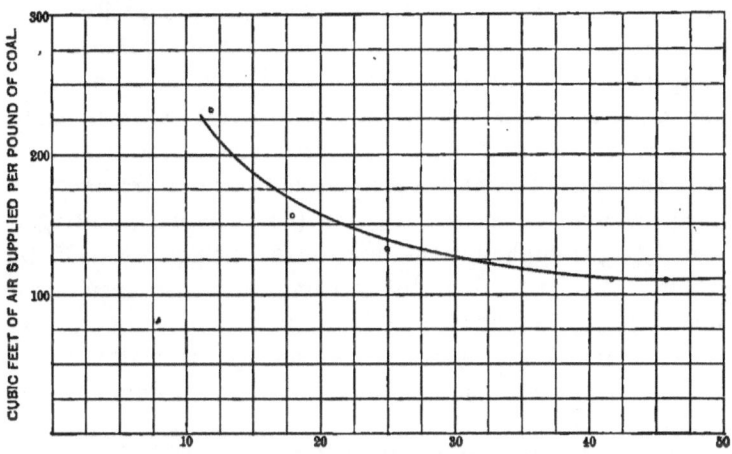

Fig. 11. AIR SUPPLY WITH DIFFERENT RATES OF COMBUSTION.

excess of air up to a combustion of 30 pounds of dry coal per hour per square foot of grate. An evaporative efficiency of 11.69 pounds of water from and at 212° per pound of combustible was recorded when burning 45.4 pounds, while the table shows that there was an actual deficiency in the air-

---

* "Experiments with Automatic Mechanical Stokers." J. M. Whitham. Transactions American Society of Mechanical Engineers, vol. XVII.

supply. The natural explanation of maintained efficiency with this air-supply is to be found in the construction of the stoker. It consists of a number of hollow grate-bars, to the ends of which the air is admitted and through numerous small holes in the upper surfaces of which it escapes. The air is thus thoroughly diffused throughout the fire; there is practically perfect contact of air and fuel, and consequent combustion with the minimum of waste gases.

"Doubts may be entertained," so states Hoadley,* "as to so large excess of air as 150 per cent occurring in practice. In fact, it is very common. It is not easy to carry on complete combustion by means of natural draft with less than 100 per cent excess, *i.e.*, double the necessary quantity. . . . Experiments to ascertain the composition, volume, and temperature of the gases from 17 boilers, burning good anthracite coal at a known rate, with great care and under most favorable conditions of draft, grate-area, rate of combustion, area of heating surface, and general arrangement, gave, by analysis, carbon dioxide (no monoxide), nitrogen, and free atmospheric air—the latter being one half the whole. A check upon the accuracy of these results was found in the temperature of the furnace. . . . In my opinion, it is understating rather than overstating the matter to say that the average of good practice would show a double supply of air."

The deductions of Clark, already given in the preceding chapter, regarding the relation of grate-area, heating-surface, water, and fuel, serve to confirm the statement that an increased rate of combustion does not entail decreased efficiency when the total consumption remains constant. Regardless of any reduction in the air-supply required per pound of coal when the rate is increased, he showed by the results of extended experiment † upon locomotive-boilers that the

---

* "Warm-blast Steam-boiler Furnace." J. C. Hoadley.
† "The Steam-engine." D. K. Clark.

efficiency remained practically constant when the rate of combustion increased and the total evaporation proceeded in the ratio of the square of the surface-ratio. The results, with values calculated therefrom, are presented in Table No. 87.

TABLE No. 87.

EFFECTS OF DIFFERENT SURFACE-RATIOS AND RATES OF COMBUSTION.

Designation Groups of Tests.	Coke Consumed per Square Foot of Grate per Hour. Pounds.	Average Ratio of Heating-surface to Grate-surface.	Coke Consumed per Square Foot of Heating-surface per Hour. Pounds.	Water Evaporated per Square Foot of Heating-surface per Hour. Pounds.	Water Evaporated per Pound of Coke. Pounds.
A	42.7	52	0.82	7.4	9.0
B	55.0	66	0.83	7.6	9.1
C	86.0	72	1.19	10.6	8.9
D	126.0	90	1.40	12.5	8.92

It is to be noted that under the experimental conditions the capacity of the group of boilers, as measured in pounds of water evaporated per hour per square foot of heating-surface, increased from 7.4 to 12.5, or 69 per cent, while the efficiency remained substantially constant. It is a perfectly reasonable inference that had the capacity been maintained constant, the efficiency would have increased with the rate of combustion.

Further and more direct evidence that with a given total coal-consumption the efficiency of the boiler rises as the rate of combustion is increased, is presented by the tests of Burnat* upon a French boiler with grates of three different areas. The principal items in these results are given in Table No. 88, which shows that with practically the same quantity of coal consumed per hour on the three grates the efficiency increased as the grate-area was diminished in the ratios of 7.26 pounds, 7.54 pounds, and 7.79 pounds of water per pound of coal, although in the third case a larger supply of air was provided.

---

* Bulletin de la Société Industrielle de Mulhouse, vol. xxx.

TABLE No. 88.

RESULTS OF PERFORMANCE OF FRENCH BOILER WITH GRATES OF DIFFERENT AREAS.

Area of Grate. Square Feet.	Air at 62° Fahr. per Pound of Coal. Cubic Feet.	Average Temperature.		Coal Consumed.		Residue. Per cent.	Water per Pound of Coal from and at 212°. Pounds.
		Feed-water. Degrees.	Gas at Dampers. Degrees.	Per Hour. Pounds.	Per Hour per Square Foot of Grate. Pounds.		
24.70	161	124	576	125	5.28	16.5	7.26
12.37	164	124	612	127	11.00	18.7	7.54
9.03	180	117	570	124	14.74	19.0	7.79

Similar confirmatory results * with grates 6 feet and 4 feet long are presented in Table No. 89, showing that higher efficiency and rapidity of evaporation were obtained from the shorter grate with about equal quantities of coal per hour.

TABLE No. 89.

EFFECT OF LENGTH OF GRATE UPON EFFICIENCY AND RAPIDITY OF EVAPORATION.

Items.	Length of Grate.	
	4 Feet.	6 Feet.
Total coal per hour.........................pounds	400	414
Coal per hour per square foot of grate........ pounds	14	23
Water at 212° evaporated per pound of coal.... pounds	10.10	10.91

The inference must of necessity be drawn from the preceding that an increase in the surface-ratio and a corresponding increase in the rate of combustion will under proper conditions result in raising the efficiency. Under the present conditions of boiler-design and the arrangements existing in most boiler-plants, the simplest means of securing the desired results would appear to lie in the introduction of feed-water, or air-

---

* "The Steam-engine." D. K. Clark.

heaters, or similar devices of such proportions that the gases resulting from the more rapid combustion will be prevented from escaping at too high a temperature.

Thickness of Fire.—It is commonly accepted that to economically burn an increased quantity of coal per square foot per hour it is necessary to increase the thickness of the layer of fuel. Comparative tests with different thicknesses of fire in a marine boiler, by Richardson and Fletcher,* showed, by Table No. 90, that the efficiency increased with the thickness of the fire.

TABLE No. 90.

EFFICIENCY OF THICK FIRES.

Items.	Thickness of Fire.		
	9-inch.	12-inch.	14-inch.
Coal consumed per square foot of grate per hour..................................pounds	27	27	27
Water evaporated per pound of coal, as supplied at 212°............................pounds	10.77	11.23	11.54

There are conditions, however, where the thinner fire may prove the more economical, as where the rate of combustion is such that a heavy fire fed at long intervals and given but little attention is compared with a thinner fire fed more frequently and run under better management. Evidently, stronger draft is required with the thicker fire; but this, as has already been pointed out, should be an element in the increased efficiency because of the greater pressure which causes more intimate contact of air and fuel.

* Report on the Boiler and Smoke-prevention Trials conducted at Wigan, 1869.

CHAPTER VIII.

DRAFT.

Definition.—Air is a necessity to the combustion of all fuel. Not only must it be supplied in volume sufficient to meet the chemical requirements, but force must be exerted to compel it to pass through the fuel and thereby cause it to come in contact with the incandescent carbon. This force is measured by the pressure-difference which causes the flow and overcomes the resistances, and is commonly designated as the draft. As usually employed, this term applies to the difference in pressure between the external air and the gases as they leave the boiler, although, as related to the combustion of the fuel, it should properly apply to the difference between the under- and over-grate pressures. It is generally measured in inches of water by means of a draft-gauge. The term "draft" is sometimes, though seldom, employed as a measure of the volume or weight of the gases passing through the fire. As the readings of a draft-gauge give no direct indication of their volume, the quantity of air or gases must be determined by other means.

In the case of a chimney, the maximum intensity of draft exists only with the maximum temperature of the gases; but after the temperature reaches about 600° F. their density decreases more rapidly than their velocity increases, so that the weight of air supplied is a maximum at about this temperature. As the draft is almost universally measured by the difference in pressure, expressed either in inches of water or in weight per unit of area, the term will be here employed as indicating the intensity or force with which it acts. This difference in pressure, whether it be the result of creating a plenum in the ash-pit or a partial vacuum in the boiler-furnace, is always necessary to produce the flow of air through the fuel whereby combustion is maintained. It is evident, therefore, that the draft or pressure-difference and the velocity or air-flow are interdependent.

Relation of Pressure and Velocity.—As the laws which govern the movement of gases are the same as those which apply to liquids in motion, their application can be most readily illustrated by means of a liquid, which has visible sub-

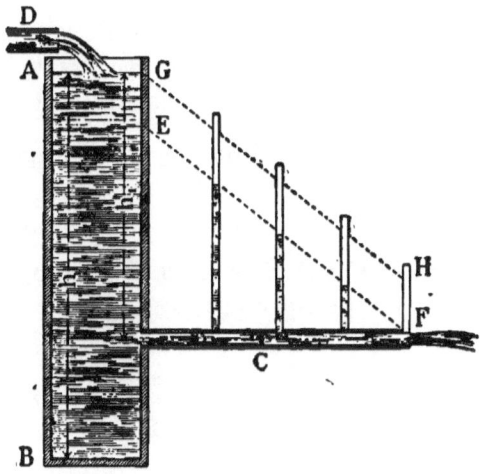

FIG. 12.

stance. If a vessel with vertical sides, as indicated in Fig. 12, be filled with water at 50° F. to the level A, the total pressure upon the bottom will be equal to the weight of the entire quantity of water. If the area of the base be 100 square inches, and the total weight of water be 1500 pounds, the pressure per square inch will be $\frac{1500}{100} = 15$ pounds. This indicates that each column of water having 1 square inch for its base, and the distance $AB = h$ for its height, weighs 15 pounds. As the weight of water per cubic foot at 50° is

62.409 pounds, and consequently 0.0361 per cubic inch, it is also evident that the distance AB must be $\dfrac{15}{0.0361} = 415.3$ inches $= 34.6$ feet. This depth of water, h, producing the given pressure per square inch, is known as the total head, and in this case is also the hydrostatic, or pressure, head. Obviously the pressure exerted is directly proportional to the head or depth of water. For, the cross-section of the vessel remaining constant, any change in the depth of the water must result in a coincident change in the total weight of water which presses upon the bottom.

If, now, a pipe C be inserted in the side of the vessel at such a height above the base that the distance h_1 is 25 feet, it is evident that the pressure per square inch of cross-section of the vessel at this point will be $\dfrac{25}{34.6}$ of that upon the base; that is, $\dfrac{25}{34.6} \times 15 = 10.85$ pounds. In other words, it will be equivalent to the head of water multiplied by its density, as clearly shown by the equations

$$p = hd \quad \text{and} \quad h = \frac{p}{d},$$

in which $p =$ pressure, $h =$ head, $d =$ density.

This pressure is transmitted to the water in the end of the pipe at its junction with the vessel. If, now, the vessel be arranged to receive a continual supply of water through D, such that its level A will be kept constant notwithstanding the outward flow through the pipe C, the head h_1 will continue the same, as will also the pressure which is exerted upon the water in the end of the pipe. By the action of gravity the water seeks to escape through the pipe, and its effect or pressure is exactly proportional to the head or depth h_1.

If there were no friction whatever in the pipe, the total head would be rendered effective for producing flow, and the speed or velocity of that flow would be exactly that which would finally result if a body under the action of gravity had freely fallen the distance measured by the head h. Therefore, its velocity is determinable by the well-known formula for falling bodies, viz.,

$$v = \sqrt{2gh},$$

in which $v =$ velocity in feet per second;
$h =$ the head in feet;
$g =$ acceleration due to gravity $= 32.16$.

As it has already been shown that $h = \frac{p}{d}$, this formula, as applied to the velocity of movement of fluids, may take the form

$$v = \sqrt{2g\frac{p}{d}}.$$

In practice the walls of the pipe under consideration would restrict the freedom of flow, and a portion of the total head or pressure would have to be expended in overcoming the resistance. This resistance is naturally greatest at the part of the pipe farthest removed from the outlet, for the freedom of flow increases as the water nears the end of the pipe where there is no resistance; the atmospheric pressure at this point being balanced by that upon the upper surface of the water in the vessel. If the pipe should be provided with a series of small open-topped gauges, as shown, they would respectively indicate, by the heights of water within them, the resistances which exist at the different points. The decreasing heights of these columns as they approach the end of the pipe is evidence of the decreasing resistance. The regularity of this fall is indicated by the dotted line EF, while the fall in the total head available at each of these points is represented by the line GH. Since these lines are parallel, the vertical distance between them, which represents the portion of the head utilized for producing velocity, is evidently constant.

As the total head represents the sole means of producing movement of the water, whatever portion of this head is used for overcoming resistance reduces by just so much the amount that remains available for the production of velocity. That portion of the total head which is thus employed to overcome resistance is known as the *pressure-head*, while that remaining and utilized for the production of velocity is designated as the *velocity-head*.

Evidently, if the pipe be of uniform diameter and the water be considered non-compressible, the velocity must be uniform, and hence the velocity or pressure expended for producing that velocity must also be uniform. This confirms the evidence of the parallel dotted lines in the figure.

With a constant total head any increase in the length of the pipe naturally increases the pressure-head and consequently reduces the velocity-head. If, however, the pipe be entirely removed, leaving only the orifice, the pressure-head will be practically eliminated and the total head will become the velocity-head. The actual velocity through the opening will be very slightly less than that calculated by the formula, owing to the slight friction of the water in passing through. But the volume passing through an opening in a flat plate will be considerably less than that which would be calculated by multiplying the area of the opening by the velocity, or rate of flow. This is due to the fact that the stream contracts as it leaves the opening, so that its minimum area, which is at a short distance from the orifice, becomes only about two thirds of the area of the opening. The effect of other forms of opening is such as to render necessary the determination of the coefficient of discharge for each.

Efflux of Air.—As the pressure is dependent upon both the height and the density of the fluid, it is evident that for a given pressure the less the density the greater the height of the column. But the law of falling bodies recognizes the fact that it is the distance fallen through and not the weight of the body that determines its velocity. Therefore, the less dense a body the higher the column required to produce a given pressure and the greater the velocity of discharge. From this it is evident that the velocity of a gas issuing under a given pressure would be greater than that of a liquid under the same conditions. And conversely, the more dense the fluid issuing at a given velocity the greater must have been the pressure to produce that velocity.

In the case of a liquid, the atmospheric pressure upon the inlet and outlet of a containing vessel is balanced and the actual height or head may be actually measured. But air is invisible, and there is no tangible distinction in substance between that producing the pressure and that constituting the surrounding atmosphere.

The pressure of the atmosphere is due to the weight of the air, and, for any area, is to be measured by the weight of a column of air having the given area as a base, and a height equal to that of the atmosphere. But this height cannot be accurately determined, and, furthermore, the density of the air decreases in geometric ratio as the distance from the earth increases. For the purposes of calculation, however, the practical equivalent of such a column may be determined by assuming the air to be of uniform density throughout, and the column of such a height as to weigh the same and to produce the same effective pressure per unit of area.

Under the standard conditions of barometric pressure of 29.921 inches, the atmospheric pressure is 14.69 pounds per square inch, or 2115.36 pounds per square foot. At this pressure a cubic foot of dry air at 50° has a density of 0.077884 pounds. Consequently a homogeneous column $\frac{2115.36}{0.077884} = 27,160$ feet high, having a base of one square foot area, would weigh 2115.36 pounds and exert this pressure upon the given area.

If air under this head were to be allowed to flow freely into a vacuum, the velocity, by the formula, would be

$$v = \sqrt{2g \times 27160}$$
$$= \sqrt{64.32 \times 27160}$$
$$= 1321.7 \text{ feet per second.}$$

In the case of air flowing into a vacuum the total head is actual, although here reduced for simplicity to that of a homogeneous column. But under any other conditions both the pressure-head and the velocity-head are purely ideal. The fact that a given air-pressure exists in a reservoir does not indicate the existence of an actual column of air of known density. But the height of such an ideal column is readily determinable by calculation from the simple equation

$$h = \frac{p}{d},$$

and it is this ideal height that is used in calculation.

If, as is frequently the case, the pressure is expressed in inches of water, as indicated by the balanced height of a column of that liquid in a water-gauge, this may be readily transformed into the height of the equivalent column of air. Thus, if the pressure-difference in inches of water be represented by H, and the equivalent head of air in feet by h, the value of h will be

$$h = \frac{\text{density of water} \times H}{12 \times \text{density of air}};$$

then at a temperature of 50° F.

$$h = \frac{62.409 \times H}{12 \times 0.077884}$$
$$= 66.77 H.$$

If this value be employed in the formula for velocity, it becomes

$$v = \sqrt{64.32 \times 66.77 H}$$
$$= 65.5\sqrt{H},$$

from which may be approximately determined the velocity of efflux of air under any given pressure-difference expressed in inches of water. As the value of H is dependent upon the temperature of both air and water, more particularly the former, it is evident that the value of the constant applies only under the stated conditions; and the value of v must be considered as only approximate where the formula is employed under other conditions without suitable corrections therefor.

But for more refined calculation additional factors must be taken into consideration. For simplicity the atmospheric pressure and humidity may be considered constant. In the case of air, which is compressible by pressure and expansible by heat, the density varies greatly with the pressure; and a change in the temperature has a most important influence. The effect of increased density, which may be produced by the pressure, is to decrease the ideal velocity-head. An opposite influence is exerted by the temperature, for by an increase in temperature the density is decreased. The velocity being dependent solely upon the ideal head, it is of the utmost importance that comparisons be reduced to the same conditions of pressure, temperature, and density.

If the pressure is to be expressed in ounces per square inch, which is readily reducible to inches of water, the formula

$$v = \sqrt{2g\frac{p}{d}},$$

when applied under the conditions of— .

g = acceleration due to gravity = 32.16,
p = pressure in ounces per square inch,
d = density or weight of one cubic foot of dry air at 50° temperature, and under atmospheric pressure = 0.077884 pounds,

becomes

$$v = \sqrt{64.32 \times \frac{p \times 144}{16 \times 0.077884 \times \frac{235+p}{235}}}.$$

In this form it is evident that the density varies with the pressure, for p being expressed in ounces, and the atmospheric pressure of 14.69 pounds being equivalent to 235 ounces, the density which exists at any given pressure p is $0.077884 \times \frac{235+p}{235}$.

The formula reduces to

$$v = \sqrt{64.32 \times \frac{p \times 9}{0.077884} \times \frac{235}{235+p}}$$

$$= \sqrt{\frac{1746659 \times p}{235+p}}.$$

The formula, when the pressure is expressed in inches of water, and allowance is made for the increased density of the air due to the pressure, takes the form

$$v = \sqrt{\frac{1746659 \times h}{406.7 + h}}.$$

Both formulæ thus take into account the compression of the air due to its pressure, but make no allowance for change of temperature during discharge.

In the accompanying Table No. 91 the effect of this compression of the air due to the pressure is taken into account, but no allowance is made for change of temperature during discharge. The air is assumed to be dry and of 50° F. temperature.

The work theoretically required to move air, as expressed in foot-pounds, is measured by the product of the distance moved and the total resistance which is overcome. In practice there are usually frictional losses which considerably increase this theoretical result.

TABLE No. 91.

VELOCITY CREATED WHEN AIR UNDER A GIVEN PRESSURE IN INCHES OF WATER IS ALLOWED TO ESCAPE INTO THE ATMOSPHERE.

Pressure in Inches of Water, per Square Inch.	Velocity of Dry Air at 50° Temperature Escaping into the Atmosphere through any Shaped Orifice in any Pipe or Reservoir in which the Given Pressure is Maintained.		Pressure in Inches of Water, per Square Inch.	Velocity of Dry Air at 50° Temperature Escaping into the Atmosphere through any Shaped Orifice in any Pipe or Reservoir in which the Given Pressure is Maintained.	
	In Feet per Second.	In Feet per Minute.		In Feet per Second.	In Feet per Minute.
0.1	20.72	1243.3	2.6	105.33	6320.0
0.2	29.30	1758.0	2.7	107.33	6439.7
0.3	35.84	2150.4	2.8	109.28	6557.0
0.4	41.43	2485.6	2.9	111.21	6672.3
0.5	46.31	2778.7	3.0	113.09	6785.5
0.6	50.73	3043.5	3.1	114.95	6896.8
0.7	54.78	3287.0	3.2	116.77	7006.3
0.8	58.56	3513.5	3.3	118.57	7114.1
0.9	62.10	3726.1	3.4	120.34	7220.2
1.0	65.45	3927.2	3.5	122.08	7324.7
1.1	68.64	4118.4	3.6	123.80	7427.7
1.2	71.68	4301.0	3.7	125.49	7529.3
1.3	74.60	4476.1	3.8	127.16	7629.4
1.4	77.41	4644.5	3.9	128.80	7728.2
1.5	80.12	4806.9	4.0	130.43	7825.7
1.6	82.73	4963.9	4.25	134.40	8064.1
1.7	85.27	5116.1	4.5	138.26	8295.4
1.8	87.73	5263.7	4.75	142.00	8520.1
1.9	90.12	5407.3	5.0	145.65	8738.8
2.0	92.45	5547.1	5.25	149.20	8951.8
2.1	94.72	5683.4	5.5	152.66	9159.7
2.2	96.94	5816.5	5.75	156.05	9362.8
2.3	99.11	5946.4	6.0	159.35	9561.2
2.4	101.23	6073.6	6.25	162.59	9755.4
2.5	103.30	6198.1	6.50	165.76	9945.8

Measurement of Draft.—Draft is usually measured by the difference in level of a liquid in the arms of a tube of U form having one end open to the atmosphere and the other connected with the enclosed space within which a different pressure exists. The preponderance of pressure in one arm forces the liquid downward and causes a corresponding rise in the other. The difference in level represents the height of a column of the liquid which will be sustained by the excess of pressure. For large pressure-differences, mercury is used because of the relatively small range which, owing to its great density, is required to show great differences in pressure. Water, however, is ordinarily employed in the measurement

TABLE NO. 92.

PRESSURES IN OUNCES PER SQUARE INCH CORRESPONDING TO VARIOUS HEADS OF WATER IN INCHES.

Head in Inches.	Decimal Parts of an Inch.								
	.0	.1	.2	.3	.4	.5	.6	.7	.8	.9
0	......	0.06	0.12	0.17	0.23	0.29	0.35	0.40	0.46	0.52
1	0.58	0.63	0.69	0.75	0.81	0.87	0.93	0.98	1.04	1.09
2	1.16	1.21	1.27	1.33	1.39	1.44	1.50	1.56	1.62	1.67
3	1.73	1.79	1.85	1.91	1.96	2.02	2.08	2.14	2.19	2.25
4	2.31	2.37	2.42	2.48	2.54	2.60	2.66	2.72	2.77	2.83
5	2.89	2.94	3.00	3.06	3.12	3.18	3.24	3.29	3.35	3.41
6	3.47	3.52	3.58	3.64	3.70	3.75	3.81	3.87	3.92	3.98
7	4.04	4.10	4.16	4.22	4.28	4.33	4.39	4.45	4.50	4.56
8	4.62	4.67	4.73	4.79	4.85	4.91	4.97	5.03	5.08	5.14
9	5.20	5.26	5.31	5.37	5.42	5.48	5.54	5.60	5.66	5.72

TABLE NO. 93.

HEIGHT OF WATER-COLUMN IN INCHES CORRESPONDING TO VARIOUS PRESSURES IN OUNCES PER SQUARE INCH.

Pressure in Ounces per Square Inch.	Decimal Parts of an Ounce.								
	.0	.1	.2	.3	.4	.5	.6	.7	.8	.9
0	......	0.17	0.35	0.52	0.69	0.87	1.04	1.21	1.38	1.56
1	1.73	1.90	2.08	2.25	2.42	2.60	2.77	2.94	3.11	3.29
2	3.46	3.63	3.81	3.98	4.15	4.33	4.50	4.67	4.84	5.01
3	5.19	5.36	5.54	5.71	5.88	6.06	6.23	6.40	6.57	6.75
4	6.92	7.09	7.27	7.44	7.61	7.79	7.96	8.13	8.30	8.48
5	8.65	8.82	9.00	9.17	9.34	9.52	9.69	9.86	10.03	10.21
6	10.38	10.55	10.73	10.90	11.07	11.26	11.43	11.60	11.77	11.95
7	12.11	12.28	12.46	12.63	12.80	12.97	13.15	13.32	13.49	13.67
8	13.84	14.01	14.19	14.36	14.53	14.71	14.88	15.05	15.22	15.40
9	15.57	15.74	15.92	16.09	16.26	16.45	16.62	16.79	16.96	17.14

of draft in connection with steam-boilers. One cubic foot (1728 cubic inches) of water at 50°, the temperature at which the preceding table has been calculated, weighs 62.409 pounds; therefore a column of water of this temperature 1728 inches high and 1 square inch cross-sectional area would exert a pressure of 62.409 pounds per square inch, and a pressure of 1 pound per square inch would be exerted by a column $\frac{1728}{62.409} = 27.7$ inches high. From this it is readily deduced that an ounce pressure per square inch is produced by a water-column 1.73 inches high, and that 1 inch head of water is equivalent to a pressure of 0.578 ounces per square inch. Table No. 92 serves to show these relations for different heights of water-column. Table No. 93 indicates the height of water-column corresponding to any given pressure in ounces per square inch.

A simple form of U-tube draft-gauge is shown in Fig. 13. The scale is laid out so as to indicate total differences in level, for which purpose it is necessary to have the water stand at exactly zero in both tubes when they are open to the atmosphere. For reading other pressures, a rubber tube may be slipped over the end of the left-hand tube and connected to the space within which the pressure is to be determined.

A somewhat unique and exceedingly compact form of gauge, acting upon this principle, may be constructed by making the glass tubes of considerably different diameters and inserting one within the other. The lower end of the outer tube should be permanently closed, and the inner tube held rigidly in position, with its lower end just out of contact with the bottom of the outer tube. If connection be made to the top of the inner tube, the change in level may be clearly seen in the outer tube, upon the surface of which the graduations may be placed.

FIG. 13.—WATER-GAUGE.

Gauges of the form and construction just described serve all ordinary purposes, but lack the refinement necessary to the determination of very small differences of pressure. For this purpose special devices are necessary. One of these is shown in Fig. 14. This embodies the principle of the U tube;

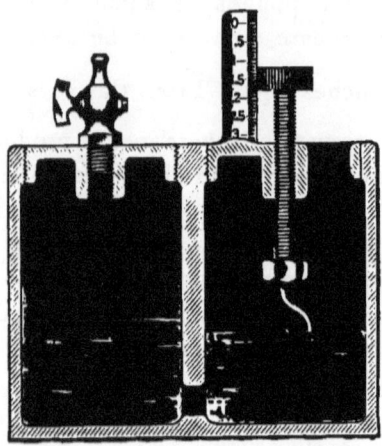

FIG. 14.—HOOK DRAFT-GAUGE.

but the level of the water is very closely ascertained by the aid of a hook-gauge. As is evident in the illustration, this instrument consists, in effect, of two tubes of relatively large diameter made in one casting. The tubes communicate near the bottom through an opening whereby a constant level of the water within them may be maintained so long as they are both exposed to the same pressure. The hook having been set so that its point is just at the surface of the water, and the micrometer-reading having been taken, connection may be made to the air-cock at the left, and the hook again adjusted at the new level. The micrometer-screw, with its graduated head, makes the reading of pressure-differences to $\frac{1}{100}$ of an inch a comparatively simple matter.

Evidently, such a gauge gives only relative readings, but may be so graduated as to indicate the total difference in water-levels. These two readings, the difference between which gives the pressure in inches of water, render it unnecessary to bring the water to any stated level under atmospheric pressure.

An extremely delicate but easily read draft-gauge is that described by Weisbach, under the name of the Wollaston Anemometer, and perfected by Hoadley and Prentiss.* As constructed and used by them, it consists of two glass tubes (Fig. 15) about 30 inches long and about 0.4 inches diameter inside, connected at each end, by means of stuffing-boxes, to suitable tubular attachments, through which they are secured to a backing of wood. A stop-cock in each of these attachments serves to establish or shut off communication between the glass tubes. Connecting with the top of each is a brass drum 4.25 inches in diameter, with heads of glass. Each drum is provided with a nipple and stop-cock for connection by tube to any desired space. Two sliding scales are provided between the glass tubes, to measure, one the depressions, and the other the elevations, of the liquid filling the lower half of the tubes.

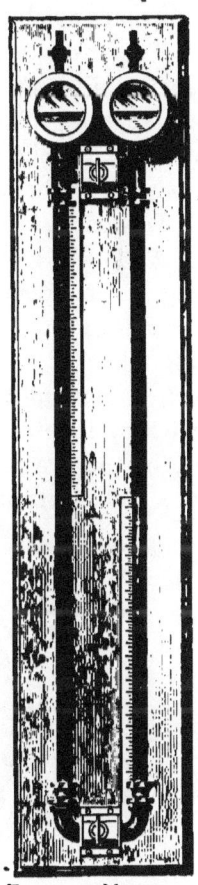

FIG. 15.—MODIFIED WOLLASTON ANEMOMETER.

The lower stop being open, the two tubes are filled up to about the middle of their heights with a mixture of alcohol and water. The lower stop-cock is then closed, the upper one opened, and crude olive-oil is carefully poured in until it fills the first tube up to the upper cross-tube, whence it flows into the second tube, and so finally fills both tubes and rises to the middle of both drums.

The oil forms, with the water-and-alcohol mixture, a very

---

* " Warm-blast Steam-boiler Furnace." J. C. Hoadley.

fine meniscus, which is readily discernible because of the contrast in colors. The specific gravity of oil should differ from that of the mixture by at least 1 per cent to avoid the tendency to mix. For most purposes, a difference of 2 per cent in the specific gravity will give sufficient sensitiveness—fifty times as much as a water-column.

This instrument, with the respective specific gravities of 0.976 and 0.937, was sensitive enough, as applied by Hoadley, to show plainly the reduction of chimney-draft caused by opening a sliding register in the fire-door for the admission of air above the fire, although the aggregate area of the opening was only six square inches.

All of the gauges which have been described are designed only for independent observations, so that an approach to a continuous record can only be secured by a multitude of readings taken at very short intervals. The impracticability of such a method points to the advantages of an instrument which by its own operation records the changes in the intensity of the draft. Such is the draft-recorder shown in Fig. 16. This instrument consists of two essential parts. First,

FIG. 16.—DRAFT-RECORDER.

the cylindrical chamber in which is contained a rubber diaphragm which moves under the influence of the draft. The motion of this diaphragm is, like that of a steam-engine indicator, multiplied by the attached arm, which carries at its end a reservoir containing ink. The second essential portion is the dial or chart, which is usually graduated so as to indicate the pressure or vacuum in inches of water. This chart, which is of paper, is held in place upon a circular plate which is caused to revolve by a system of clockwork. The point of the ink reservoir, being kept elastically in contact with the revolving dial, continuously records all variations in the draft.

In measuring the pressure exerted by moving air, both the velocity-head and the pressure-head have to be taken into account. To separate these two factors of the total head, a form of Pitot's tube may be employed, as illustrated in Fig. 17, where it is applied in connection with a pipe, through the side of which it is inserted. The tube A is open at the end and connects by rubber tubing with one arm of an ordinary U-tube water-gauge. The other tube, B, is closed upon the end, but has in its opposite sides two small holes, and is connected to the other arm of the gauge. Tube A receives the full effect of the current of moving air, and thus tends to indicate upon the gauge the total head, including both the velocity-head and the pressure-head.

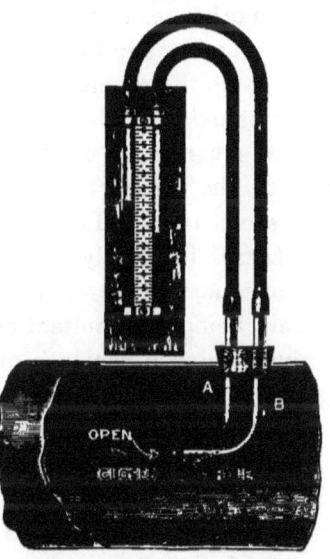

FIG. 17.—PITOT'S TUBE.

But the influence of the velocity is practically removed from B, which, therefore, receives only the pressure due to the pressure-head. As this tube is connected to the other arm of the gauge, the pressure thus indicated is only that due to the velocity-head; for, both arms being subject to the pressure-head, these pressures are balanced.

At high pressures even this device is not altogether reliable, for the air moving by the openings in tube B has an aspirating influence which may tend to produce a partial vacuum in this tube. It is, therefore, necessary, before making final measurements, to determine by independent readings whether this is the case, and to what extent.

Conditions of Boiler-draft.—In boiler-practice the force of the draft must be expended in two ways. First, a portion of it is necessary to overcome the resistances of the grate and the fuel upon it, of the combustion-chamber, flues or tubes, and uptake, and of the means of connection to the source of draft, be it fan or chimney. Within the chimney or fan certain other resistances must also be overcome. Second, the draft must in addition be sufficient to impart the necessary velocity to the requisite amount of air for the direct purposes of combustion. The velocity thus produced varies directly as the square root of the intensity of the draft, and consequently the volume at constant temperature likewise varies in the same ratio. The force expended in overcoming the resistances is directly proportional to the pressure,—that is, to the square of the velocity,—while the work done in moving the air, being the resultant of a given pressure exerted through a given distance, which is measured by the velocity, becomes proportional to the product of these factors; namely, to the cube of the velocity. Therefore, if under stated conditions the resistance be increased, as by a thicker fire or finer coal, the intensity of draft required for overcoming this resistance must also be increased. If, however, the prior conditions were such that the maximum intensity of draft attainable was already devoted to the requirements of combustion, any demand for increased draft to overcome the greater resistance could only be met by reducing the amount of that portion previously devoted to the creation of velocity. But a slight reduction in the velocity considerably decreases the expenditure necessary for overcoming the resistance. The ultimate result, however, is that the expenditure for overcoming resistance is increased, while the velocity and consequent volume of air are decreased. But both of these changes are proportionately less than would at first appear. The reason is evident in the fact that, whereas the resistances which are overcome are directly proportional to the pressure or draft exerted, the velocity is decreased only in proportion to the square root of the draft, which is thus diverted from its service of producing velocity to that of overcoming the added resistance.

In all boiler-practice the most important of the resistances to be overcome are those of the tubes and of the grate with the fuel upon it, while the expenditure for the production of velocity is comparatively small. A careful study of these resistances is, therefore, of importance in a thorough consideration of the practical conditions of draft production.

Gale* found, in the case of an ordinary stationary boiler-furnace, the following pressures in pounds per square foot:

Required to produce entrance-velocity (3.6 feet per second)......... 0.013
Required to overcome resistance of fire-grate...................... 0.91
Required to overcome resistance of combustion chamber and boiler-tubes... 1.23
Required to overcome resistance in horizontal flue................. 0.06
Required to produce discharge-velocity (11.2 feet per second)...... 0.085
Total effective draft-pressure..................................... 2.298
Back-pressure due to friction in stack............................. 0.19
Total static pressure produced by chimney.......................... 2.488

This total pressure, given in pounds per square foot, is equivalent to 0.28 ounces per square inch, or 0.48 inches of water.

It is evident from this table that the greater part of the draft-pressure is, as already stated, necessary to overcome the resistances presented by the fuel and the boiler-tubes. In fact, Rankine asserts that the throttling action at the grate is ordinarily sufficient to cause a loss of head equal to about three-quarters of the whole draft of a chimney. As about 75 per cent of this remaining fourth is necessary to balance the frictional resistances of the flues and the chimney, there remains only about one-sixteenth of the total head for the production of velocity. Gale† cites an instance in which he found

---

* "Theory and Design of Chimneys." Horace B. Gale. Transactions American Society of Mechanical Engineers, vol. XI.

† Transactions American Society of Mechanical Engineers, vol. XI. p. 777.

three-quarters of the whole draft of a chimney. As about 75 per cent of this remaining fourth is necessary to balance the frictional resistances of the flues and the chimney, there remains only about one-sixteenth of the total head for the production of velocity. Gale * cites an instance in which he found that about 60 per cent of the entire head was lost by throttling at the grate, and only about 4 per cent of the total head was actually expended in accelerating the gases. In this case the height of the chimney was 92 feet, and the temperature of the gases 609°. If the whole head had been employed in producing velocity of the gases, their velocity, as calculated by Gale, would have been about 78 feet per second. In reality the mean observed velocity was only 16 feet per second.

Although the ultimate object of any means of draft production must necessarily be to create draft or velocity sufficient to provide the required amount of air and to carry off the gases, yet this portion of its work is almost infinitesimal as compared with the demand made for sufficient pressure to overcome the resistance of the fuel and the boiler. In other words, the ability to create sufficient pressure-difference is the primary requisite to burning a given quantity of fuel, rather than the ability to move a certain amount of air. Draft-producing apparatus is not, therefore, to be based merely upon the total number of cubic feet to be moved per hour, as determined by multiplying the coal-consumption by the allowance of air per pound of coal. If this were the case, a low chimney, or a large slow-running fan, would meet the requirements. In reality, the relatively immense resistances of fuel and boiler demand that the chimney or fan shall first be designed to create sufficient intensity of draft to overcome these resistances and to create the requisite velocity. This velocity must be such that, if multiplied by the full area at

---

* Transactions American Society of Mechanical Engineers, vol. XI. p. 777.

words, the ability to create sufficient pressure-difference is the primary requisite to burning a given quantity of fuel, rather than the ability to move a certain amount of air. Draft-producing apparatus is not, therefore, to be based merely upon the total number of cubic feet to be moved per hour, as determined by multiplying the coal-consumption by the allow-

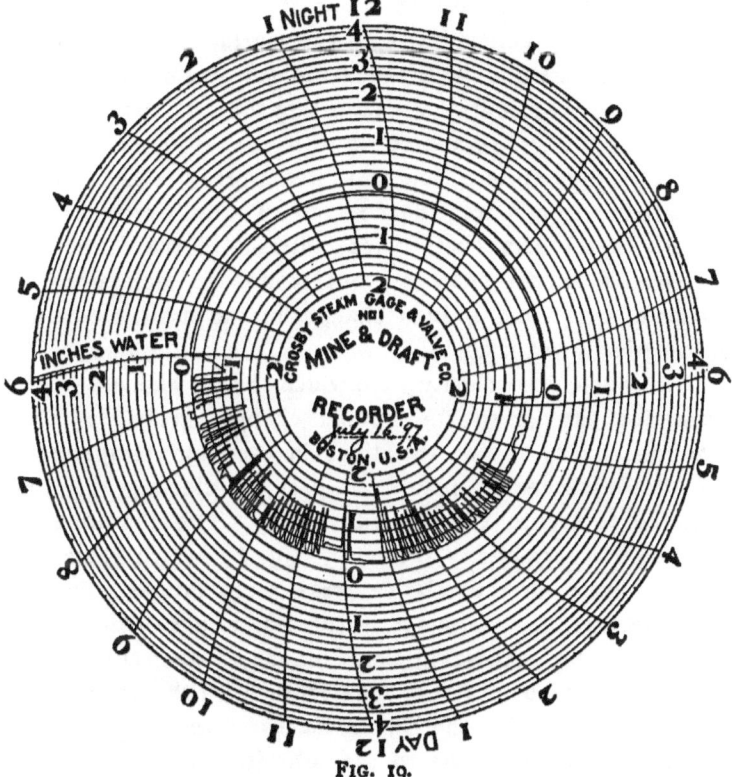

FIG. 19.

ance of air per pound of coal. If this were the case, a low chimney, or a large slow-running fan, would meet the requirements. In reality, the relatively immense resistances of fuel and boiler demand that the chimney or fan shall first be designed to create sufficient intensity of draft to overcome these resistances and to create the requisite velocity. This velocity must be such that, if multiplied by the full area at which it is measured, the product will equal the volume of air necessary for the combustion of the stated amount of fuel. The height of chimney or the diameter and speed of fan necessary to create the draft thus shown to be required having been determined, it is only necessary to make the capacity such as to accommodate the given volume of air.

Constant steam-pressure is one of the desirable elements in good boiler-practice, but with the varying conditions in the state of the fire and in the demand for steam this can only be maintained by continually adjusting the draft to meet these conditions. Figs. 18 and 19 serve to make this clear. The former is the reproduction of a steam-pressure chart, while the latter is from a coincident record of a draft-recorder connected to the same battery of boilers. In the former the pressure is almost constant, but it was so maintained by means of great and sudden changes in the draft, which was automatically regulated.

Of course the velocity with which the air and gases pass from the ash-pit to the uptake changes greatly as they progress, owing to the variations in area and temperature. All of these changes play their part in affecting the draft required to secure the desired results. The conditions in practice are very clearly shown by Fig. 20, which represents in graphical form the relative areas, temperatures, volumes, and velocities, as determined in the case of one of the boilers of the International Navigation Company's steamship Berlin. This boiler was one of four connecting with the same stack. It was provided with the Ellis & Eaves system of preheating the air before admission to the ash-pit, and the necessary draft was produced by four special Sturtevant steam-fans, through which the gases were drawn and thence discharged into the funnel. The relative actual observed temperatures are indicated in diagram 2, together with a statement of the portion of the passage to which they pertain. The first diagram graphically represents the area existing at the various points, and the third indicates the corresponding absolute

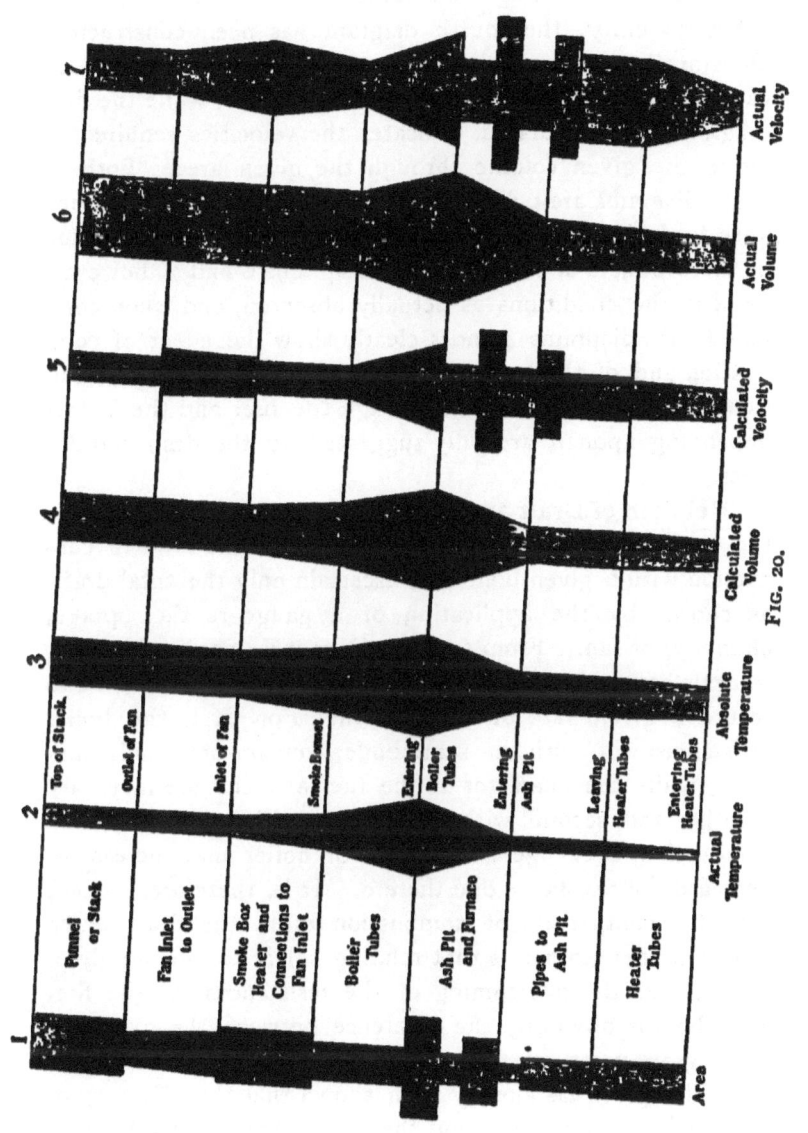

Fig. 20.

temperatures. Taking the volume of air entering the heater-tubes as unity, the fourth diagram has been constructed, showing the relative volumes to which the air would in each case be expanded by the existing temperature; while the fifth diagram, also calculated, indicates the velocities required to move the given volume through the given area. Both of these diagrams are purely theoretical, for they make no allowance for leakage or for increased volume due to the accession of the products of combustion. Diagrams 6 and 7, however, present the conditions as actually observed, and when compared with diagrams 4 and 5 clearly show the effect of combustion and of the leakage, which, owing to existing conditions, was large. The area through the fuel and the factors depending upon it are only suggested by the dash and dot lines.

Relation of Draft and Rate of Combustion.—It has been the usual practice, in the determination of the draft in connection with a given boiler, to ascertain only the total draft, as shown by the application of a gauge to the uptake, chimney, or fan. From such readings it is manifestly unfair to draw conclusions as to the amount or intensity of draft to secure a given rate of combustion. For, first, the boiler resistances will, with the same boiler, remain practically constant, while the character of the fuel and the fire may vary greatly; and, second, with the same conditions as to fuel and combustion, a change in the type of boiler may increase or decrease the resistance due thereto. It is, therefore, evident that for comparison of combustion rates, the draft which should be determined is that relating solely to the supply of air to, and the overcoming of the resistances in, the fire. This draft is obviously the difference between the over- and under-grate pressures.

Although it has already been shown that the efficiency of combustion may increase and the required air-supply decrease as the combustion rate rises, for the sake of simplicity in the matter of comparison the required air-supply per pound of coal may here be taken as constant for all rates of combustion. The volume of air (for constant temperature) then becomes an index of its velocity, and varies as the square root of the effective pressure or draft. Conversely, the required draft will vary as the square of the rate of combustion. Of course, for the purpose of properly proportioning draft-producing apparatus, the resistances of all parts of the boiler, including the fuel upon the grate, should be ascertained for all ordinary coals, types of boiler, and conditions of draft.

The difficulties in the way of obtaining such knowledge are evident. It is not easy to secure identical conditions of boiler and draft when testing different coals, and even with the same coal there may be variations in its size, in the manner in which it is fired, and in which the grates are kept clear of ashes, which very seriously affect the results. As indicative of the variation in draft-pressure required for different kinds of coal, rates, and stages of combustion, the results in Table No. 94 are presented. These are from a test * of a

TABLE No. 94.

DRAFT CONDITIONS WITH COXE STOKER.

Items.	Size of Coal.		
	Buckwheat.	Buckwheat.	Rice.
Pressure in igniting compartment. ...in. of water	0.14	0.25	0.44
Pressure in burning compartment.... " " "	0.31	0.56	0.89
Pressure in burning-down compartment " " "	0.24	0.49	0.73
Pressure in burning-out compartment.. " " "	0.17	0.42	0.67
Pressure of blast of air, average.. ..... " " "	0.24	0.43	0.68
Vacuum in furnace.................. " " "	0.10	0.15	0.24
Total furnace-draft.................. " " "	0.34	0.58	0 92
Vacuum in stack-flue................ " " "	0.13	0.40	0.58
Total draft .... " " "	0.37	0.83	1.26
Pounds of dry coal per hour per sq. ft. of grate....	19.8	32.9	28.0

Coxe stoker applied to Babcock & Wilcox boilers. This stoker, which is of the travelling chain-grate type, with the fire upon its upper surface, is provided with four blast com-

---

* " Experiments with Automatic Mechanical Stokers." J. M. Whitham Transactions of American Society of Mechanical Engineers, vol. XVII.

partments under the fire. To each of these air is admitted in the desired proportion from a supply furnished by a fan.

Whitham, in the same paper, also presents a table given here as Table No. 95, showing the relation of size of coal to results obtained with a Wilkinson stoker, from which the increased draft required by small sizes of coal is made evident. It is here that mechanical draft becomes most beneficial in making possible the combustion of low-grade (because finely divided) fuel.

TABLE No. 95.

RELATIVE RATES OF COMBUSTION OF SMALL SIZES OF ANTHRACITE COAL.

Grade of Coal.	Size of Coal (Round Holes, Punched Plates(.	Relative Rates of Combustion for Same Draft.
Pea................	Through $\frac{7}{8}$ inch and over 9/16 inch	100
Buckwheat..........	Through 9/16 inch and over $\frac{3}{8}$ inch	85
Rice................	Through $\frac{3}{8}$ inch and over 3/16 inch	70

The relation of the size of the coal to the intensity of the draft required for its combustion is most forcibly shown by the results determined by Coxe,* and here presented in Table No. 96.

TABLE No. 96.

RESULTS OF TESTS OF PEA AND BUCKWHEAT COAL.

Kind of Coal.	Rate of Combustion per Square Foot of Grate per Hour.	Pounds of Water Evaporated from and at 212° per lb. of Coal.	Air-pressure in Inches of Water.	Maximum Limit to Size of Coal in Inches.
Oneida pea coal.............	13.63	8.56	0.375	$\frac{7}{8}$
" No. 1 buckwheat....	13.58	7.94	0.5	9/16
" No. 2 " ...:	11.40	8.60	0.625	$\frac{3}{8}$
" No. 3 "	11.34	8.65	1.04	3/16
Eckley No. 3 "	9.44	8.75	1.125	3/16

* "Furnace for Burning Small Anthracite Coals." Eckley B. Coxe. Trans. Am. Inst. Mining Engineers, vol. XXII.

It will be noted that the smallest coal requires over three times the draft to secure a combustion rate less than 70 per cent of that obtained with the largest size.

Present methods of boiler-testing lack the refinements in ascertaining the draft-pressure at different points which are necessary to an intelligent comparison of results. This fact is emphasized by the results given in Table No. 97,* which

TABLE No. 97.

RELATION OF AIR-PRESSURE AND INDICATED HORSE-POWER PER SQUARE FOOT OF GRATE.

Name of Ship.	Mean Air-pressure in Fire-rooms. Inches.	Mean I. H. P. per Square Foot of Grate-surface.
Orlando	1.02	16.17
Undaunted	1.87	16.17
Australia	1.77	17.75
Galatea	1.14	18.50
Narcissus	1.02	16.17
Immortalitè	2.01	17.93

presents the observed conditions of draft and mean indicated horse-power per square foot of grate, in the trial tests of a number of steam vessels almost identical in their steam-power equipment. Draft was in each case produced by fans discharging into a closed fire-room. The fact that under similar conditions the fire-room pressures varied from 1.02 to 2.01 inches is, at least, sufficient to emphasize the necessity of readings which indicate the effective pressure-differences between ash-pit and furnace-chamber.

Under practically identical conditions of boiler, coal, and firing, there is still opportunity, particularly with mechanical draft, for divergence from any established relation, owing to a probable decrease in air-volume per pound of coal as the rate of combustion increases, and to a coincident increase in

* "Artificial Draft and its Effects on Boiler-construction." E. Lechner. Translated by Asst. Engr. Emil Theiss, U. S. Navy. *Journal of American Society of Naval Engineers*, Aug., 1891.

the thickness of the fire. These tend, however, to counteract each other. A greater depth of fuel on the grates naturally indicates a proportionately greater resistance. But the rate of combustion always increases at a more rapid rate than the thickness of the fire; therefore, the pressure under which the greater volume of air would be supplied would also increase more rapidly than the thickness of the fire, and hence more readily tend to overcome this resistance.

The combined effect of all the factors appears to be to bring the rate of combustion to substantially the theoretical basis first presented; viz., proportional to the square root of the effective pressure. Thus, for instance, two similar tests upon the same boiler in the works of the B. F. Sturtevant Co., Boston, Mass., gave results presented in Table No. 98. The square roots of the effective pressures in the two cases are, respectively, 0.705 and 0.542, which are in the ratio of 1 to 0.769, while the rates of combustion are in the ratio of 1 to 0.767.

TABLE No. 98.

RELATION OF DRAFT AND RATE OF COMBUSTION.

Designation of Test.	Vacuum in Furnace. Inches.	Vacuum in Ash-pit. Inches.	Effective Pressure to Produce Combustion. Inches.	Coal Burned per per Hour per Sq. Ft. of Grate. Pounds.
A	0.640	0.144	0.496	21.45
B	0.372	0.078	0.294	16.45

The results of a series of tests of the locomotive-boiler of a torpedo boat, as presented by E. Lechner[*] in a discussion of this subject, are given in Table No. 99. Tests numbered 1, 2, 3, and 4 were conducted upon the regular grate under artificial draft produced by a centrifugal fan in a closed fire-

---

[*] "Artificial Draft and its Effects on Boiler-construction." E. Lechner. Translated by Assistant Engineer Emil Theiss, U. S. Navy. *Journal of American Society of Naval Engineers*, August, 1891.

room, while in tests numbered 5, 6, and 7 the grate was reduced to one-half the area. In the first series of tests the air-supply per pound of coal was reasonable in quantity, but practically constant for different rates of combustion; while in the second series, owing to the existing circumstances, it was large in volume but decreased as the rate of combustion increased.

TABLE No. 99.

RESULTS OF TESTS OF LOCOMOTIVE-BOILER ON TORPEDO-BOAT WITH DIFFERENT RATES OF COMBUSTION.

Number of Test.	Grate-surface. Sq. Ft.	Heating-surface. Sq. Ft.	Ratio of Heating to Grate-surface.	Gauge-pressure. Pounds per Sq. In.	Air-pressure in Inches of Water.		
					Fire-room.	Furnace.	Chimney.
1	20.44	882.32	43.1	113.6	1.97	1.57	0.39
2	....	....	....	....	2.95	2.36	0.59
3	....	....	....	....	3.94	3.15	0.91
4	....	....	....	....	5.90	4.53	1.18
5	10.22	882.32	86.2	113.6	2.95	2.17	0.39
6	....	....	....	....	5.90	3.94	1.00
7	....	....	....	....	6.80	4.53	1.18

Number of Test.	Coal per Hour per Sq. Foot of Grate. Pounds.	Water Evaporated per Hour from 86°. Pounds.	Water Evaporated per Pound of Coal. Pounds.	Temperature of Chimney-gases. Degrees.	Air Supplied per Minute. Cubic Feet.	Air per Pound of Coal. Cubic Feet.	Mean Thickness of Fuel on Grates. Inches.
1	53.6	437.8	8.16	518	4112	225.2	12.60
2	63.4	517.5	8.17	608	5365	248.4	13.40
3	76.4	583.6	7.64	626	6287	241.6	14.17
4	93.4	649.8	7.00	716	7092	222.9	15.35
5	66.7	528.6	7.93	554	5221	459.6	9.84
6	101.4	710.0	7.00	608	6831	395.5	11.82
7	113.0	770.2	6.80	698	7092	368.4	13.78

The relation existing between the rate of combustion and the square root of the corresponding pressure-difference is presented in Table No. 100. A comparison of columns 4 and 5, or of columns 6 and 7, indicates that the rate of combustion advances in approximately the same proportion as the square root of the pressure-difference, taking each series

STEAM-BOILER PRACTICE.

Table No. 100.

RELATION OF SQUARE ROOT OF PRESSURE-DIFFERENCE TO RATE OF COMBUSTION.

Number of Test.	Difference of Pressure between Fire-room and Furnace. Inches.	Square Root of Difference of Pressure between Fire-room and Furnace.	Ratio of Square Roots of Difference of Pressure referred to Test No. 1 as Unity.	Ratio of Square Roots of Difference of Pressure referred to Test No. 5 as Unity.	Ratio of Corresponding Rate of Combustion, referred to Test No. 1 as Unity.	Ratio of Corresponding Rate of Combustion, referred to Test No. 5 as Unity.
1	0.40	0.632	1.00	....	1.00	....
2	0.59	0.768	1.21	....	1.18	....
3	0.79	0.888	1.40	....	1.42	....
4	1.37	1.170	1.85	....	1.74	....
5	0.78	0.883	....	1.00	....	1.00
6	1.96	1.400	....	1.59	....	1.52
7	2.27	1.507	....	1.71	....	1.70

of tests by itself. An attempt to compare the second series with the first results in a series of values which do not continue the ratio of the first series, but which, if multiplied by a constant quantity, may be brought into accord therewith. The approximate relation between rate and pressure is thus indicated, as well as the fact that a complete formula, which shall enable one to determine the pressure-difference required for a given rate of combustion for any kind of fuel, must include a constant which shall apply only under the given conditions, and whose various values for all conditions can only be determined by experiment.

Such a formula would naturally take the form

$$w = c \sqrt{p_1 - p_2},$$

in which w = pounds of coal burned;
p_1 = pressure in ash-pit;
p_2 = pressure in furnace-chamber;
c = constant dependent on type of boiler, kind of fuel, depth of fire, etc.

But until the value of c is determined for all conditions, the formula must be limited in its application. In a word, the relation between the *effective* pressures and different rates of combustion under the same conditions is substantially established. But the exact draft or pressure-difference required to maintain a given rate of combustion of a specific kind of coal in a particular type of boiler can only be approximated in the present state of knowledge.

The relation between the *total* draft and the rate of combustion, which is that usually indicated in the results of an ordinary test, is practically all that is at present available. But this relation is not of necessity the same as that between the *effective* pressure and the rate. In fact, the constant character of the resistances of the boiler proper and the variableness in those of the fuel with different rates cause this relation to depart somewhat from that holding in the case of the effective pressures.

This is clearly indicated in the results of careful tests of Whitham,[*] in which the draft was taken at various points in the passage of the air and gases through the furnace and boiler when coal was being burned at different rates of combustion. The boiler was of the horizontal return tubular type, 60 inches diameter by 20 feet long, with 44 4-inch tubes. The grate was stationary, 5 feet by 5 feet 4 inches, with 46 per cent air-space, the ratio of heating-surface to grate-surface being 42.6. In certain tests the tubes were fitted with retarders made of strips of No. 10 iron 20 feet long, with a pitch of 10 feet. Cambria Company coal (run-of-mine, bituminous) was used. The recorded drafts and rates of combustion are given in Table No. 101. The results are also graphically presented in Fig. 21, in which "fair" lines are drawn through the various points plotted. There is also added a line based upon the relation already stated,—that the rate of combustion should vary as the square root of the effective draft. Taking the furnace draft as representative of this effective draft, and

---

[*] "The Effect of Retarders in Fire-tubes of Steam-boilers." J. M. Whitham. Transactions American Society of Mechanical Engineers, vol. XVII.

224   STEAM-BOILER PRACTICE.

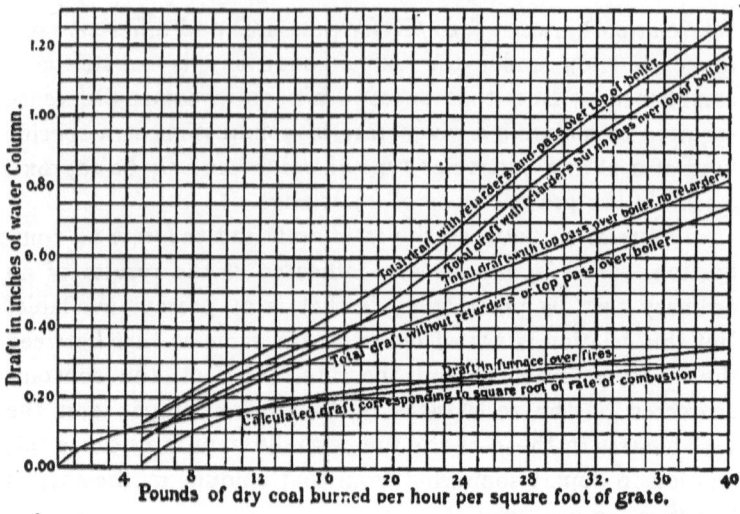

Fig. 21.

TABLE No. 101.
RELATION OF DRAFT AND RATE OF COMBUSTION.

Pounds of Dry Coal burned per Hour per Square Foot of Grate.	Furnace Draft.	Resistance, in Inches of Water, Due to			Total Draft in Inches of Water.			
		Pass under Boiler and through Tubes. No Retarders.	Retarders in Tubes.	Pass over top of Boiler.	No Retarders. No Return Pass.	With Retarders. No Return Pass.	Top Pass. No Retarders.	With Top Pass and Retarders.
5	0.04	0.04	0.00	0.04	0.08	0.08	0.12	0.12
8	0.11	0.05	0.02	0.04	0.16	0.18	0.20	0.22
10	0.13	0.07	0.03	0.05	0.20	0.23	0.25	0.28
12	0.17	0.07	0.04	0.05	0.24	0.28	0.29	0.33
14	0.19	0.10	0.03	0.05	0.29	0.32	0.34	0.37
15	0.20	0.11	0.03	0.05	0.31	0.34	0.36	0.39
16	0.21	0.12	0.03	0.05	0.33	0.36	0.38	0.41
18	0.23	0.13	0.06	0.05	0.36	0.42	0.42	0.48
20	0.24	0.16	0.08	0.06	0.40	0.48	0.46	0.54
22	0.26	0.18	0.12	0.06	0.44	0.56	0.50	0.62
25	0.27	0.22	0.19	0.06	0.49	0.68	0.55	0.74
28	0.29	0.24	0.27	0.07	0.53	0.80	0.60	0.87
30	0.30	0.27	0.31	0.07	0.57	0.88	0.64	0.95
34	0.32	0.31	0.38	0.08	0.63	1.01	0.71	1.09
36	0.33	0.34	0.40	0.08	0.67	1.07	0.75	1.15
40	0.36	0.38	0.46	0.08	0.74	1.20	0.82	1.28

starting at a rate of 12 pounds as unity, the corresponding theoretical drafts have been calculated and plotted for other rates. These correspond with those for the furnace draft, except at very low rates of combustion, but diverge decidedly from the lines representative of other elements of the total draft.

The total draft required for the efficient combustion of various kinds of fuels, as given by Hutton,* is presented in Table No. 102. Of course this must be considered as apply-

TABLE No. 102.

TOTAL DRAFT REQUIRED FOR EFFICIENT COMBUSTION OF DIFFERENT KINDS OF FUEL.

Kind of Fuel.	Total Draft in Inches of Water.	Kind of Fuel.	Total Draft in Inches of Water.
Straw	0.20	Slack, very small	0.7 to 1.1
Wood	0.30	Coal dust	0.8 to 1.1
Sawdust	0.35	Semi-anthracite coal	0.9 to 1.2
Peat, light	0.40	Mixture of breeze and slack	1.0 to 1.3
Peat, heavy	0.50	Anthracite, round	1.2 to 1.4
Sawdust mixed with small coal	0.60	Mixture of breeze and coal dust	1.2 to 1.5
Steam coal, round	0.4 to 0.7	Anthracite slack	1.3 to 1.8
Slack, ordinary	0.6 to 0.9		

ing only under ordinary conditions, and as approximate at the best, for he does not even give the rate of combustion. Still further discussion of this subject is presented in the succeeding chapter.

Leakage of Air.—A fruitful source of poor draft and decreased efficiency lies in the leakage of air through boiler settings. The extent of such infiltration is frequently surprising, being often so great that the flame of a match is drawn to and into the interstices of an 8-inch brick wall, not alone at fine visible cracks, but at mortar joints apparently sound. Evidently, the result of such leakage with suction draft is to

---

* "Steam-boiler Construction." Walter S. Hutton.

increase the volume of air to be handled and to decrease the temperature; thereby inevitably reducing the draft in the case of a chimney, but in the case of a fan of proper size fortunately tending to increase the suction, unless the volume be in excess of the capacity of the fan. This results because, with a fan at constant speed, the intensity of the draft increases as the temperature of the gases passing through it is reduced. Coincident with the reduction of temperature is an increase in the weight of the gases—and hence of the admitted air—handled by the fan without change in speed. With forced draft beneath the grates the opposite tendency is noticeable, and a more or less direct loss of heat is the result. Under any condition the leakage naturally increases with the draft-pressure, no matter how produced, and under equal pressures is obviously no greater with mechanical than with chimney draft.

An indication of the frequent amount of such leakage is shown in the results of the chemical analyses of samples of gas taken respectively from the back end of a boiler just before entering the tubes and from the uptake flue, as presented in Table No. 103. These results are sufficient to show the important necessity of preventing such loss.

TABLE No. 103.

AIR-LEAKAGE THROUGH BOILER-SETTINGS.

Conditions.	Coal consumed per Square Foot of Grate per Hour. Pounds.	Percentage of Air which leaked in between Back End and Uptake Flue.
Middle boiler in operation, dampers on other two boilers closed and packed....	21.45	22.08
Conditions the same...................	16.45	27.59
Three boilers in operation..............	15.09	15.40

The plant consisted of three boilers in one battery. In the first two tests an inward movement of air was perceptible at the doors of the outside boilers not in use, although the uptake dampers had been very carefully packed. The decreased leakage in the third case is principally due to the reduction of the ratio between exposed surface of the setting and the volume of air passing through the uptake flue. The original introduction of sheet-metal stops in the setting of such boilers is a comparatively simple matter, and if carefully carried out practically prevents all leakage. The lack of such arrangements, however, tends, in many instances, to render misleading the results of tests in which allowance has not been made for the effects of the incidental leakage.

CHAPTER IX.

CHIMNEY DRAFT.

Principles of Chimney Draft.—If two chimneys of identical construction and dimensions be connected at the bottom by a passage of the same cross-sectional area, and one of them be provided at its base with a means of heating the air, a definite air movement will result as soon as heat is applied. The cold air in one chimney, being heavier than the heated air in the other, will constantly seek to secure equilibrium of weights and pressures by flowing downward to the base of the heated chimney. As a natural consequence, the heated air will be forced upward and the cold air which takes its place will, in turn, be heated and follow the same course. A continuous flow will thus be maintained, its velocity and consequent volume being dependent upon the difference in density of the two columns of air; that is, upon the pressure difference. Although the difference in density results from the application of heat, the air movement is purely mechanical in its character, and depends directly upon the action of gravity.

The total difference in pressure upon the internal bases of the two chimneys is exactly equal to the difference in weight of the two columns of air within them. The relative difference may be expressed in any convenient terms, as pounds per square foot, ounces per square inch, or by the height of a column of water, mercury, or other fluid necessary to balance this pressure. If a simple U-tube pressure-gauge be partly filled with water, one end connected to the base of the cold chimney, and the other to the base of the hot chimney, the preponderance of weight of the air in the former will force the water downward in that arm of the tube and cause a corresponding rise in level in the other arm. The total difference in level may be read in inches of water and then readily resolved into ounces per square inch, or pounds per square foot.

Evidently, there being no difference between the character of the air in the cold chimney and that in the surrounding atmosphere, the same relative pressures will exist, and the same flow will continue if the cold chimney be removed and the air be allowed to directly enter the base of the hot chimney. Furthermore, the relative differences in the density and pressure created, being measured respectively by unit volume and unit area, are independent of the cross-sectional area of the hot chimney. In other words, the pressure-difference is dependent only upon the height of the chimney and the difference in density between the heated air within and the cold air without. The terms " hot " and " cold " are, of course, only relative, for the draft is primarily dependent upon the actual temperature-difference.

That changes in the temperature, either of the external atmosphere or the gases within the chimney, have a most marked influence upon the draft is very clearly shown by Table No. 104, in which the draft, as indicated in inches of water, is given for a chimney 100 feet high, with various internal and external temperatures. For any other height of chimney than 100 feet the height of the water column is directly proportional to that of the chimney. Hence, doubling the height doubles the draft. This is not to be confused with the fact that the velocity which the draft has the power to create and the corresponding volume of air moved vary as the square root of the height. This table clearly indicates the necessity of high chimney temperatures for ample draft, and readily accounts for the stronger draft which exists in cold weather because of the greater temperature-difference.

TABLE No. 104.

HEIGHT OF WATER-COLUMN DUE TO UNBALANCED PRESSURES IN CHIMNEY 100 FEET HIGH.

Temperature in Chimney.	Temperature of External Air.										
	0°	10°	20°	30°	40°	50°	60°	70°	80°	90°	100°
200°	.453	.419	.384	.353	.321	.292	.263	.234	.209	.182	.157
220	.488	.453	.419	.388	.355	.326	.298	.269	.244	.217	.192
240	.520	.488	.451	.421	.388	.359	.330	.301	.276	.250	.225
260	.555	.528	.484	.453	.420	.392	.363	.334	.309	.282	.257
280	.584	.549	.515	.482	.451	.422	.394	.365	.340	.313	.288
300	.611	.576	.541	.511	.478	.449	.420	.392	.367	.340	.315
320	.637	.603	.568	.538	.505	.476	.447	.419	.394	.367	.342
340	.662	.638	.593	.563	.530	.501	.472	.443	.419	.392	.367
360	.687	.653	.618	.588	.555	.526	.497	.468	.444	.417	.392
380	.710	.676	.641	.611	.578	.549	.520	.492	.467	.440	.415
400	.732	.697	.662	.632	.598	.570	.541	.513	.488	.461	.436
420	.753	.718	.684	.653	.620	.591	.563	.534	.509	.482	.457
440	.774	.739	.705	.674	.641	.612	.584	.555	.530	.503	.478
460	.793	.758	.724	.694	.660	.632	.603	.574	.549	.522	.497
480	.810	.776	.741	.710	.678	.649	.620	.591	.566	.540	.515
500	.829	.791	.760	.730	.697	.669	.639	.610	.586	.559	.534

For a full comprehension and application of the principles of chimney draft it is necessary to consider them mathematically. If h be the height of the chimney, d the density of the external air, and d_1 that of the heated air, the pressure-difference p for unit area may be expressed as

$$p = hd - hd_1 = h(d - d_1).$$

The height of a column of external air which would produce this pressure, acting simply by its weight, may be found by dividing the pressure by the density of the external air. Therefore, if H represents the height of such a column, the expression will be

$$H = \frac{p}{d} = \frac{d - d_1}{d}.$$

The theoretical velocity with which the external air would

enter the chimney, if no resistance existed, may be expressed by the equation

$$v = \sqrt{2gH};$$

$$= \sqrt{2gh\left(\frac{d-d_1}{d}\right)};$$

in which v = velocity, in feet per second;
g = the acceleration due to gravity = 32.16;
H = head, or distance fallen, in feet.

The preceding deductions are based upon the assumption that there is no resistance to the movement of the air. Such a condition evidently cannot exist in the operation of the ordinary chimney. The motion of the gases creates a certain back pressure due to their friction on the inner surface of the chimney. This back pressure must be deducted from that due to the difference in density in order to ascertain the effective pressure which may be applied to compel the air to pass through a boiler-furnace and up the chimney.

The effective pressure which in any given case is necessary to overcome the frictional and other resistances of the air in its passage through the fuel, the boiler-tubes and other portions of the boiler, and impart to the gases the necessary velocity of entrance and exit, must depend upon the existing conditions of character of fuel, design and proportions of boiler, and other considerations which render exact determination extremely difficult.

Chimney Design.—From the preceding it is obvious that a chimney must be so designed as to create sufficient draft or pressure-difference to overcome all resistances, and in addition impart the necessary velocity to the required amount of air. The number of pounds of coal which can be burned in a given time on a given grate equals the weight of air forced through, divided by the number of pounds of air required for the combustion of a pound of coal under the given conditions. As already shown, this latter amount may vary all the way from that theoretically necessary up to an amount 100 per cent or more in excess thereof. The weight of the air forced through in the given period of time is equal to the area through which it is admitted, multiplied by its velocity and by its density or weight per unit of volume. To impart this velocity, and also that through the furnace, boiler-tubes, smoke-flue, and chimney, there is necessary a pressure which may be approximately ascertained by calculations based upon theoretical considerations. The additional pressure required to overcome the various resistances does not, however, admit of theoretical determination, and can only be found by direct experiment, as has already been pointed out. These resistances are proportional to the square of the actual velocity, and depend upon the diameter and length of tubes, flues, and chimney, the thickness of the fuel and its state of division. It is, therefore, obvious that the proper design of a chimney to meet given conditions must be based upon the results of experiments under similar conditions. This fact readily accounts for the great divergence in the formulæ which have been presented, and for the necessity of a series of constants practically determined for application under stated conditions.

The earlier formulæ deduced by Péclet, Rankine, Morin, and Weisbach have until recently been generally accepted, but the theory upon which they are based has of late been the object of considerable criticism. Péclet represented the law of draft by the formula

$$h = \frac{u^2}{2g}\left(1 + G \times \frac{fl}{m}\right),$$

in which $h =$ "head" or height of hot gases which, if added to the column of gases in the chimney, would produce the same pressure at the furnace as a column of outside air, of the same area of base, and height equal to that of the chimney;

$u =$ required velocity of gases in chimney;

$G =$ a constant to represent resistance to the passage of air through the coal;
$l =$ length of the flues and chimney;
$m =$ mean hydraulic depth, or the area of cross-section divided by the perimeter;
$f =$ a constant depending upon the nature of the surfaces over which the gases pass, whether smooth, or sooty and rough.

Rankine's and the other formulæ are somewhat similar in form. The impossibility of assigning proper values to the constants in these formulæ has, up to the present time, prevented their practical application for chimney design, and resort has, as a consequence, been made in most cases to empirical methods.

Based upon this theory, however, the usual formula for determining the number of pounds of coal which may be burned per hour, under the conditions of the draft which may be created by a given chimney, takes the simple and practical form of

$$F = CA\sqrt{H},$$

in which $F =$ number of pounds burned per hour;
$A =$ area of cross-section of chimney;
$H =$ height of chimney;
$C =$ coefficient varying from 10 to 20, according to the conditions.

As it stands, the formula implies that it is a matter of indifference in regard to the draft whether, for instance,—other things remaining the same,—a chimney is built 81 feet high and 2 feet square inside, or 16 feet high and 3 feet square inside; for in either case the product $A\sqrt{H}$ is the same. Common sense at once proves the impracticability of such a formula thus broadly applied. It is, therefore, usually qualified by limiting its application to cases when the total grate-area is about eight times that of the chimney. But evidently such limitation is arbitrary, for this ratio of grate- to chimney-area is not constant for all plants.

The empirical formulæ of Kent,* which have been very generally adopted in the design of chimneys, are based upon the observed conditions in a large number of boiler-plants, and take the general primary form of

$$A = \frac{0.06F}{\sqrt{h}};$$

$$h = \left(\frac{0.06F}{A}\right)^2;$$

in which A = area of chimney;
h = height of chimney;
F = pounds of coal burned per hour.

The basis data are:

1. The draft power of the chimney varies as the square root of the height.

2. The retarding of the ascending gases by friction may be considered as equivalent to a diminution of the area of the chimney, or to a lining of the chimney by a layer of gas which has no velocity. The thickness of this lining is assumed to be 2 inches for all chimneys, or the diminution of area equal to the perimeter multiplied by 2 inches (neglecting the overlapping of the corners of the lining). For simplifying calculation, the coefficient is taken as the same for square and round chimneys, making the effective area E as expressed by the equation,

$$E = A - 0.6\sqrt{A};$$

in which A = total area.

3. The power varies directly as this effective area E.
4. A chimney should be proportioned so as to be capable

---

* Transactions American Society of Mechanical Engineers, vol. VI., and " Mechanical Engineers' Handbook," William Kent.

of giving sufficient draft to cause the boiler to develop much more than its rated power, in case of emergencies, or to cause the combustion of 5 pounds of fuel per rated horse-power of boiler per hour.

5. The power of the chimney varying directly as the effective area E, and as the square root of the height H, the formula for horse-power of boiler for a given size chimney will take the form, H.P. $= CE\sqrt{H}$, in which C is a constant. Adopting the general value for C of 3.33, as determined from the results of numerous examples in practice, the formula for horse-power becomes

$$H.P. = 3.33(A - .6\sqrt{A})\sqrt{H};$$

from which the values of E and H may also be obtained. In proportioning chimneys by these formulæ the height is generally first assumed, with due consideration of the conditions, and then the area for the assumed height and horse-power is calculated. The results of calculation for all ordinary dimensions of chimney and ranges of power are presented in Table No. 105.

The capacity in horse-power of a given chimney is inversely proportional to the amount of coal required per horse-power. Therefore, in large plants, where the economy of coal-consumption is high, the capacities given in the table will be proportionately increased.

Gale,* having satisfied himself, both by experiment and mathematical analysis of the incorrectness of the common theory of chimney draft upon certain points, substitutes a theory in which the height is dependent upon the stack temperature and the rate of combustion, and whose practical application is based upon certain experimentally-determined constants.

He divides the problem of chimney design into two parts:

---

* "Theory and Design of Chimneys." Horace B. Gale. Transactions American Society of Mechanical Engineers, vol. XI.

Table No. 105.

CAPACITY IN HORSE-POWER OF CHIMNEYS FOR STEAM-BOILERS.

Diameter in Inches.	Side of Equiv. Square in Inches.	Effective Area $E = A - 0.6\sqrt{A}$ in Square Feet.	Height of Chimney in Feet.													
			50	60	70	80	90	100	110	125	150	175	200	225	250	300
18	16	0.97	23	25	27	29										
21	19	1.47	35	38	41	44	66									
24	22	2.08	49	54	58	62	88									
27	24	2.78	65	72	78	83										
30	27	3.58	84	92	100	107	113	119	156							
33	30	4.48		115	125	133	141	149	191							
36	32	5.47		141	152	163	173	182	229	204						
39	35	6.57			183	196	208	219		245						
42	36	7.76			216	231	245	258	271	289	316					
48	43	10.44				311	330	348	365	389	426					
54	48	13.51					427	449	472	503	551	595				
60	54	16.98					536	565	593	632	692	748				
66	59	20.83						694	728	776	849	918	981			
72	64	25.08						835	876	934	1023	1105	1181	1253		
78	70	29.73							1038	1107	1212	1310	1400	1485	1565	
84	75	34.76							1214	1294	1418	1531	1637	1736	1830	
90	80	40.19								1496	1639	1770	1893	2008	2116	2005
96	86	46.01								1712	1876	2027	2167	2298	2423	2318
102	91	52.23								1944	2130	2300	2459	2609	2750	2654
108	96	58.83								2090	2399	2592	2771	2939	3098	3012
																3393
114	101	65.83									2685	2900	3100	3288	3466	3797
120	107	73.22									2986	3226	3448	3657	3855	4223
132	117	89.18									3637	3929	4200	4455	4696	5144
144	128	106.72									4352	4701	5026	5331	5618	6155

For pounds of coal burned per hour for any given size of chimney, multiply the figures in the table by 5.

" First, that of ascertaining the draft-pressure necessary to burn the desired quantity of fuel in the furnace; second, that of determining the dimensions of a chimney which will produce the required draft at the least expense." The mean velocity, v, with which the air enters the various openings to the furnace he shows to be equivalent to

$$v = \frac{T_a BF}{140400a},$$

in which T_a = absolute temperature of outer air;
B = number of pounds of air supplied per pound of fuel;
F = number of pounds burned per hour;
a = total area of openings through which air is admitted to the fire;

when the relative density of the external air and the gases at the same temperature is in the proportion of 40 to 39. As the effective pressure varies as the square root of the corresponding velocity, the actual pressure difference, or frictional resistance P, is expressed by

$$P = Kv^2,$$

in which K = aggregate coefficient of resistance of fire-grate, bed of coals, combustion-chamber, boiler-tubes, flue, and chimney.

Gale has, by somewhat limited experiments, established the value of K as approximately 0.2 for average conditions, but this value varies greatly in different cases.

If in the expression $P = Kv^2$ the value of v (already given) be substituted, the equation becomes

$$P = K\left(\frac{T_a BF}{140400a}\right)^2,$$

which indicates the reduction of pressure in pounds per square foot required at the bottom of the chimney which is to burn F pounds of fuel per hour, in a furnace having an area for air admission a (in square feet), and a coefficient of resistance K, allowing B pounds of air per pound of fuel.

The chimney-height, H, to produce this draft-pressure is shown to be

$$H = \frac{KT_{,}}{T_{,} - 533 - \frac{M}{A^{2}}\left(\frac{T_{,}F}{15000}\right)^{2}}\left(\frac{F}{3.5a}\right)^{2},$$

in which $T_{,}$ = absolute temperature of chimney-gases;
M = inside perimeter of chimney, in feet;
A = area of cross-section of chimney, in square feet;

and the other designations remain as in the preceding formulæ.

In this formula allowance is made for difference in density of external air and chimney-gases, and in the form of cross-section. The temperature of the external air is assumed at 59°, and the most economical proportions of a chimney taken to be such that its cost is proportional to the $\frac{1}{8}$ power of the area multiplied by the $\frac{3}{8}$ power of the height.

From the last formula is ultimately derived the expression

$$A = 0.07F^{\frac{1}{2}}$$

as an approximate formula for determining the most economical area for a chimney which is to burn F pounds of coal per hour. For chimneys of ordinary proportions the sectional area in square feet should by this formula be approximately equal to the number of pounds of fuel to be burned per minute.

For those cases in which the temperature of the chimney-gases is between 150° and 600° the formula for height may be reduced to

$$H = 100\frac{K}{t}\left(\frac{F}{a}\right)^{2},$$

in which $t =$ common temperature of chimney-gases above the zero of the Fahrenheit thermometer.

This formula may be still further simplified by adopting a value for a equal to one third of the grate-area, making $K = 0.2$, and letting G represent the area of the grate in square feet, whereby

$$H = \frac{180}{t}\left(\frac{F}{G}\right)^2.$$

In this formula the height of chimney is proportional to the square of the rate of combustion per square foot of grate divided by the common temperature of the chimney-gas. It is to be considered, however, as only a rough approximation. The value of K above must of necessity be largely affected by the character of the fuel, and further experiment is necessary to determine its value under given conditions.

The difficulties in the way of a purely theoretical consideration of the subject of chimney-draft have been pointedly considered by Kent,[*] who concludes with these words: " In the present state of our knowledge I do not think any satisfactory theory of chimneys can be framed which will include a consideration of all the different variables that affect the rate of combustion, and from which a formula can be derived that will prove of value in practice. . . . The best chimney formula that can be obtained at present is an empirical one, which may be modified or divided into two or more to meet different practical conditions, as new data are obtained from experiment."

The preceding discussion is sufficient to show the difference in formulæ and results, but is evidence that they would be in substantial accord if the constants could be based upon sufficiently exhaustive tests on the amount of grate, fuel, and boiler resistances under all conditions of practice and qualities, kinds, and conditions of fuel.

---

[*] Transactions American Society of Mechanical Engineers, vol. XI. p. 995.

Evidently, the latter factor, that of the fuel, as has already been shown in the discussion of draft, is the most important of all, and any formula which does not make proper allowance for variation in it is liable to give uncertain results. Kent's constant is based upon an assumption of fair practice, which reduces to 15 pounds of coal consumed per square foot of grate per hour, under the draft created by a chimney 80 feet high. But the character of the coal is not specified.

Trowbridge,* basing his calculation on somewhat different data, determined the average rates of combustion for various heights of chimney, as here presented in Table No. 106. The air-supply per pound of fuel was assumed at 250 cubic feet for all rates, and no attempt was made to indicate the

TABLE No. 106.

HEIGHT OF CHIMNEY TO PRODUCE CERTAIN RATE OF COMBUSTION.

Height of Chimney in Feet.	Pounds of Coal Burned per Hour per Square Foot of Cross-section of Chimney.	Pounds of Coal Burned per Hour per Square Foot of Grate, the Ratio of Grate to Cross-section of Chimney being 8 to 1.
20	60	7.5
25	68	8.5
30	76	9.5
35	84	10.5
40	93	11.6
45	99	12.4
50	105	13.1
55	111	13.8
60	116	14.5
65	121	15.1
70	126	15.8
75	131	16.4
80	135	16.9
85	139	17.4
90	144	18.0
95	148	18.5
100	152	19.0
105	156	19.5
110	160	20.0

* "Heat and Heat-engines." W. P. Trowbridge.

kind of fuel or the effect of any change in its character. It will be noticed that this table gives a rate of 16.9 pounds for an 80-foot chimney, as compared with Kent's figure of 15 pounds.

Thurston * gives this rough rule for the case of anthracite coal: "Subtract 1 from twice the square root of the height, and the result is the rate of combustion." This also gives the rate for an 80-foot chimney as 16.9.

As changes in the character of the fuel are likely to be made at any time in the life of a boiler-plant, it is obvious that the chimney-height, which is a measure of the draft-pressure, should be made sufficient to meet all requirements. How much effect the kind of fuel may have upon the height of chimney necessary for its combustion is evidenced in the results of extensive tests with telescopic stacks by de Kinder. He found that, for substantially equivalent results, the height for free-burning bituminous coals should be 75 feet, for slow-burning bituminous 115 feet, and for fine anthracite coals 125 to 150 feet.

These results point most clearly to the necessity, as already stated, of first determining the height of the chimney necessary to burn a given kind of fuel at a stated rate per square foot of grate, and then making it of sufficient area to burn the requisite amount. Evidently, in the light of common experience, as plainly shown by de Kinder's tests, the first cost and the fixed charge for the necessary chimney will vary with the kind of coal used. The existing relation may be approximately illustrated by a consideration of Table No. 105. It is there indicated that 245 horse-power of boilers may be served either by a chimney 90 feet high and 42 inches in diameter, or by one 125 feet high and 39 inches in diameter. Such differences are well within the limits for different kinds of coal, as given above. If, then, in one case the coal can be burned with the draft produced by a 90-foot

---

* "Manual of Steam-boilers." R. H. Thurston.

chimney, while in another a different kind of coal requires a chimney 125 feet high, the difference in the cost of the two chimneys must enter as an important item in the question of ultimate economy.

Upon the basis established by Gale, and previously referred to, the cost of a chimney is nearly proportional to the $\frac{1}{6}$ power of its area multiplied by the $\frac{3}{2}$ power of the height. As the area is proportional to the square of the diameter, the relative costs in the case of these chimneys become—

Cost of 90-foot chimney $= 42^{\frac{1}{3}} \times 90^{\frac{3}{2}} = 10,316$.
Cost of 125-foot chimney $= 39^{\frac{1}{3}} \times 125^{\frac{3}{2}} = 16,072$.

That is, the chimney would cost about 56 per cent more in one case than in the other to secure the same result in boiler-power.

As regards the capacity of a chimney of a given height, which is directly dependent upon its area, it is necessary to make it originally sufficient for all future probabilities. If this capacity be made too great for present conditions, it becomes necessary to reduce the volume by means of dampers. But by no means, in the case of a chimney, can the draft itself, as measured by pressure-difference and power to overcome resistances, be increased above that normally due to its height with given temperature conditions.

Thus, for instance, if with a given chimney the grate-area be reduced and the rate of combustion proportionally increased so as to maintain the same total consumption, the resistances will be increased because of thicker fires and decreased free area through them. To overcome these resistances there is demanded greater draft; but as the draft of a chimney is absolutely limited by its height, any further expenditure of its draft for this purpose must by just so much reduce the portion of its draft available for producing the requisite air-flow. If the chimney be large for the boiler-plant, and if under the lower combustion-rate it be necessary to operate it with practically closed dampers, there may be sufficient reserve to meet the requirements. But under ordinary conditions the chimney has but little power to respond to such requirements, or even to a material increase of the rate of combustion upon the regular grates. That is, its ability and capacity are distinctly limited, and to provide against contingencies it is usually necessary to build and pay the fixed charges on a structure more expensive than is regularly required.

Efficiency of Chimneys.—The chimney as a means of creating a movement of air depends upon the heating of that air; although, as shown, its actual movement is to be considered upon purely mechanical grounds. The heat thus employed is, however, absolutely wasted, so far as its utilization for any other purpose is concerned. Any attempt to extract more of the heat from the gases as they escape from the boiler must result in a reduction of the draft. This inherent loss is, therefore, always chargeable to any plant in which the draft is produced by a chimney, and possibilities in the way of increased economy must relate only to other losses so long as a given chimney is retained.

The percentage of the total calorific value of coal which is carried off by the products of combustion, and therefore available only for the production of draft, has already been presented in Table No. 47 for different degrees of excess of air and of temperature above the atmosphere. As there shown, this loss actually amounts to 19 per cent when the gases are at 500° and the excess of air is 100 per cent. Evidently such a great loss as is thus possible should require energetic effort to secure its reduction by a more economical substitute for the chimney.

Of course the weight of air moved by means of a chimney of given height must depend upon its area. As heat is the means by which this air-movement is brought about, the efficiency of the chimney must be measured by the amount of heat expended for this purpose. Heat being transformable into work, the efficiency is, therefore, to be measured by the number of foot-pounds of work represented by the pressure-difference exerted through the distance represented by the height of the column of cold air necessary to produce the given pressure, as compared with the number of foot-pounds represented by the total amount of heat expended.

Suppose, for the purposes of illustration, a chimney 100 feet high, having a cross-sectional area of 10 square feet, the atmospheric temperature at 62°, and the temperature of the chimney-gases at 500°; and further, for simplicity, assume that no work is lost in friction and that heated air is substituted for the hot gases, for their density and specific heat are approximately the same. Under these conditions the density d at 62° will be 0.0761 pounds per cubic foot, and the density d', 0.0414 at 500°. Therefore, by the formula previously given, the pressure-difference with the chimney 100 feet high will be

$$p = h(d - d')$$
$$= 100(0.0761 - 0.0414)$$
$$= 3.47 \text{ pounds per square foot.}$$

This makes the total pressure-difference $3.47 \times 10 = 34.7$ pounds over the entire area of the chimney.

The height of a column of external air which will produce the above pressure per square foot is

$$H = h\left(\frac{d - d'}{d}\right)$$
$$= 100\left(\frac{0.0761 - 0.0414}{0.0761}\right)$$
$$= 45.6 \text{ feet.}$$

The velocity of the air entering the base of the chimney under this head is

$$v = \sqrt{2gH} = \sqrt{64.32 \times 45.6}$$
$$= 54.2 \text{ feet per second;}$$

and its weight per second,

$$\text{Weight} = 54.2 \times 10 \times 0.0761 = 41.25 \text{ pounds.}$$

The movement of this air is the result of heating it from 62° to 500°; that is, through $500 - 62 = 438°$. As the specific heat of air under constant pressure is 0.2375, the total heat expended per second in moving 41.25 pounds is?

$$\text{Heat expended} = 41.25 \times 438 \times 0.2375$$
$$= 4291.0 \text{ B. T. U.}$$

As one heat-unit is equivalent to 778 foot-pounds of work, the work equivalent to the total amount of heat expended, and which goes to waste without performing useful work in heating the water, is?

$$\text{Work equivalent of heat} = 4291.0 \times 778$$
$$= 3{,}338{,}398 \text{ foot-pounds.}$$

But the work actually done is the result of overcoming a total pressure of 34.7 pounds through a distance of 54.2 feet; that is,

$$\text{Work actually done} = 34.7 \times 54.2 = 1880.7 \text{ foot-pounds.}$$

Therefore, the efficiency of the chimney is?

$$\text{Efficiency} = \frac{1880.7}{3338398} = 0.000563.$$

That is, less than six ten-thousandths of the heat expended is represented by the work done. In practice the resistance of the chimney, the cooling of the gases in their passage up it, and other causes combine to decrease even this extremely low efficiency.

If in the place of the chimney there be substituted a fan of proper size, arranged to be driven by a direct-connected engine, the efficiency with which it would move the above-stated volume of air under the given conditions, but without the chimney, may be calculated with reasonable accuracy.

Evidently the efficiency of a fan thus applied is the resultant of the efficiencies of the steam-boiler, the engine, and the fan, together with the loss by friction in the apparatus. If the combined efficiency of the boiler and engine be taken as one tenth, the efficiency of the fan at the low value of only five tenths, and the loss from friction as two tenths, or the efficiency as regards friction eight tenths, the resulting efficiency of the system will be—

$$\text{Efficiency} = 0.1 \times 0.5 \times 0.8 = 0.04.$$

That is, of the work done, or its equivalent in heat-units expended to produce a given result, one twenty-fifth is actually applied for that purpose; the remainder is lost in the processes of transformation and transmission and in friction. This efficiency, which allows for loss by friction, as was not the case with the chimney, is

$$\frac{0.04}{0.000563} = 71.05$$

times greater than that of the chimney.

It may be shown that the relative efficiency of a fan and a chimney is dependent upon the height of the chimney and not upon the difference in temperature, and that it varies inversely as the height of the chimney. Thus in the case of a chimney 75 feet high the fan would, upon the same basis as above, have an actual efficiency of

$$\frac{100 \times 71.05}{75} = 94.7$$

times greater than the theoretical efficiency of the chimney, while at 200 feet high the fan would still have 35.5 times higher efficiency; but this improvement in favor of the chimney would be largely offset by its proportionally greater cost as compared with a fan.

CHAPTER X.

MECHANICAL DRAFT.

Definition.—In its simplicity of construction and operation is to be found the reason for the continued employment of the chimney as a means of draft-production notwithstanding the excessive wastefulness of the method. But the intensity of interest in the more economical generation of steam is forcing to the front other devices by means of which the quantitative and commercial efficiency of the boiler may be increased. Among them is machanical draft.

Artificial draft, as distinguished from so-called natural draft, may be produced by means of steam-jets inducing a flow of air, by blowing-engines, by air-compressors, by positive rotary blowers, and by fan-blowers or exhausters. Although the practical success of the locomotive is largely due to Stephenson's introduction of the steam-nozzle for draft-production, it does not follow that the same method is applicable where the exhaust steam would not otherwise be wasted. The blowing-engine, the air-compressor, and the rotary positive blast-blower all possess disadvantages which render undesirable their adoption for this purpose. This narrows the field down to the steam-jet and the centrifugal fan-blower or exhauster.

Steam-jets.—The steam-jet as a means of inducing a flow of air is usually constructed upon the injector principle. It has been applied in the chimney for inducing the air-movement through the fuel, as well as in the ash-pit for forcing therein a volume of air which is caused to pass upward through the fuel. In connection with the latter arrangement, steam-jets have often been introduced to also deliver air above the fuel, frequently in the form of a number of finely divided streams designed to mix intimately with the gases arising from the fuel-bed.

The introduction of steam in conjunction with the air, which results from the use of the steam-jet, is often asserted to assist in keeping a fire free and open, particularly in the case of fine anthracite fuels. But, in so far as steam for this purpose may be necessary, it can be as well introduced in connection with a fan. Consequently, the merits of the steam-jet, as compared with a fan, must rest solely upon the relative efficiency with which a given amount of air is supplied; or, as more simply measured, by the proportion which the steam required to operate the steam-jet or fan bears to the total steam produced by the boiler in connection with which it operates. In either case the percentage of steam thus used is largely dependent upon the size of the plant, being greatest with the smallest plant.

Careful experiments, conducted at the New York Navy Yard,[*] to determine the best form of steam-jet for producing forced draft in launch-boilers, served to show the inefficiency of such devices for this purpose. The results of five series of tests are presented in Table No. 107. A different form of jet, indicated by a designating letter, was used in each case; its supply of steam being taken from a boiler separate from that to which the jet was applied. The percentages of steam are such—the maximum being 21.2 per cent—as to make the adoption of a steam-jet out of the question when any other means of draft-production can be employed. In jet C, which had a hole only one-sixteenth inch in diameter, and which was the most economical of all, the steam used was one pound in two minutes.

---

[*] Annual Report of the Chief of the Bureau of Steam-engineering, U. S. Navy, 1890.

Table No. 107.

RESULTS OF EXPERIMENTS UPON STEAM-JETS AT NEW YORK NAVY YARD.

Items.	Pounds of Water Evaporated per Hour.				
	A	B	C	D	E
In boiler making steam	463.8	580.0	361.25	528.5	545.00
In boiler supplying steam-jet	97.5	120.0	30.00	63.2	76.25
Per cent of steam made as used in steam-jet	21.20	20.70	8.30	12.00	14.00

The case of the steam-jet may be briefly summarized thus: It has the advantage of costing very little to put in and keep in repair. Its disadvantages are: first, it requires a very large amount of steam to run it; second, it introduces a large amount of water or steam, all of which has to be heated and carried up chimney; third, unless very carefully managed there is a large development of carbonic oxide, hydrogen, and marsh-gas, due to dissociation of the water, which has a tendency to carry off a great deal of heat in the stack; fourth, the intensity of draft producible by this means is distinctly limited; and fifth, the noise incident to its use is at times excessive.

Fans.—Two types of fans exist. The first, known as the disk or propeller-wheel, is constructed on the order of the screw-propeller, and moves the air in lines parallel to its axis, the blades acting upon the principle of the inclined plane. The second, or fan-blower proper, consists in its simplest form of a number of blades extending radially from the axis and presenting practically flat surfaces to the air as they revolve. By the action of the wheel the air is drawn in axially at the centre and delivered from the tips of the blades in a tangential direction. This type may be simply designated as the centrifugal fan, or, more properly, as the peripheral-discharge fan.

The propeller, or disk fan, which is available for ventilating purposes when it acts against slight resistances, is practically useless as a means of draft-production. The desired results can only be secured by the use of the peripheral-discharge type, which for this purpose is usually enclosed in a case of such shape as to provide free movement for the air as it escapes at the periphery, and an outlet through which it is all delivered.

The amount of steam required by a fan-blower employed for producing draft is, under ordinary conditions, from a fraction of 1 per cent of the total capacity of the boilers to which it is applied, up to a possible maximum of 3 or 4 per cent in small boiler-plants or with uneconomical apparatus; and practically the whole of this expenditure of power, in the form of exhaust-steam, may be subsequently utilized for heating or similar purposes. The results of the full-power forced-draft contract trials of eight vessels of the U. S. Navy show that about 0.75 per cent of the mean indicated horse-power of all the machinery was required to operate the blowers. From the necessities of such work the fans are operated at high pressures and consequently require a proportionately large amount of power for their operation.

The comparative effect of the steam-jet and the fan-blower upon the composition of the chimney-gases is well shown by a test by Coxe* upon two adjoining sets of boilers using the same fuel and fired by the same men, as presented in Table No. 108. A was fitted with a steam-jet, and B with a fan-blower and Coxe stoker. The losses indicated by the excessive presence of carbonic oxide (CO), hydrogen (H), and marsh-gas (CH_4) are noticeable in the case of the steam-jet.

The relative merits of the fan and the jet are thus expressed by Coxe: " The fan is more expensive to install and may cost more to keep in order, but where the arrange-

---

* "Some Thoughts upon the Economical Production of Steam," etc. Eckley B. Coxe. Transactions New England Cotton Manufacturers' Association, 1895.

MECHANICAL DRAFT. 251

TABLE No. 108.

COMPARATIVE GAS-ANALYSIS WITH STEAM-JET AND FAN-BLOWER.

Designa-tion.	Conditions.	Constituents.				
		O	CO	CO_2	H	CH_4
A	With steam-jet...............	0.30	13.15	8.20	11.08	2.00
B	With fan-blower...............	1.70	0.40	16.80	......	......

ments can be made to utilize the heat in the stack-gases it is more economical so far as heat-units used are concerned. It has one great advantage,—it is possible to at all times obtain the exact blast necessary to produce the best results in the furnace, which is very important."

Design of Fans.—In operation, the peripheral-discharge fan sets in motion the air within it, which, acting by centrifugal force, is delivered tangentially at the outer circumference of the wheel. Air rushes in at the axial inlet to fill the space between the blades, in which there is, by the centrifugal action, a tendency to form a vacuum. The degree of this vacuum is dependent upon the circumferential speed of the wheel; and the velocity of the air discharged, through an outlet of proper size, is substantially equal thereto.

It has already been shown that a certain ideal head is necessary to produce a given velocity. This head, which may be expressed in terms of the pressure and density, as $h = \frac{p}{d}$, is usually designated as the " head due to the velocity."

The pressure, of which this head is one of the factors, is understood to be that existing in an enclosed space from which the air escapes at a velocity expressed by the formula

$v = \sqrt{2gh} = \sqrt{2g\frac{p}{d}}$. But the pressure which this stream of moving air may exert upon any external surface with which it comes in contact may be different from that which existed in the reservoir and produced the given velocity. This external pressure is due to the impact and reaction of the air, and for a given velocity depends in quantity upon the size and form of the surface and the angle of incidence. Theoretically, if the density is considered constant, this pressure, in the case of a stream striking a plane surface at right angles, will be double that which produced the velocity; or, more clearly stated, in its relation to water, the reaction of such a stream is equal to the weight of a column of water whose cross-section is that of the stream and whose height is double that ($2h$) due to the velocity. If the plane surface be of proper dimensions and surrounded by a raised border, to prevent the ready escape of the water, the theoretical pressure will be four times the head due to the velocity. Other shapes will give other values. With air, its lack of viscosity and the partial vacuum formed on the back side of the plate influence the actual results.

In the attempt to force air at a given velocity through a given pipe, it is the province of the fan-wheel, if employed therefor, to create within the fan-case a total pressure above the atmosphere which shall be sufficient to produce the velocity and also overcome the resistances of the case and the pipe. If, however, the pipe be removed and the fan be allowed to discharge the air through a short and properly shaped outlet, the pressure necessary will, with an efficient fan, be substantially that which is required to produce the velocity.

The velocity of the fan-tips or circumference of the fan-wheel which is necessary to produce a given velocity of flow through a properly shaped outlet within the capacity of the fan is substantially equal to that velocity of flow. If, therefore, the peripheral velocity of a fan be known, the resulting pressure for the production of velocity through an outlet of proper size and shape may be calculated by the formula last given.

From this basis formula it is evident that the pressure created by a given fan varies as the square of its speed. The volume of air delivered is, however, practically constant per revolution, and therefore is directly proportional to the speed of the fan.

The work done by a fan in moving air is represented by the distance through which the total pressure is exerted in a given time. As ordinarily expressed in foot-pounds, the work per second is, therefore, the product of the velocity of the air in feet per second, the pressure in pounds per square foot, and the effective area in square feet over which the pressure is exerted. If W represents the work done, p the pressure, a the area, and v the velocity, the expression for the work becomes

$$W = pav.$$

But it has previously been shown that

$$v = \sqrt{2g\frac{p}{d}},$$

hence

$$p = \frac{dv^2}{2g}.$$

Therefore, the value of W becomes

$$W = \frac{dav^3}{2g},$$

from which it is evident that the work done varies as the cube of the velocity; that is, as the cube of the revolutions of the fan. The reason is evident in the fact that the pressure increases as the square of the velocity, while the velocity itself coincidently increases; hence the product of these two factors of the power required is indicated by the cube of the velocity.

As one horse-power is equivalent to 33,000 foot-pounds of work per minute, the horse-power for a given area of discharge in square inches, when the value of g is taken as 32.2, may be expressed by

$$\text{H.P.} = \frac{60 dav^2}{144 \times 33000 \times 64.4}$$

$$= \frac{dav^2}{5100480}.$$

Under no conditions can any device move air with the small proportionate expenditure of power indicated by this formula; for this value does not include the losses due to frictional resistances of the machine itself and of the air in its movement through the machine and connecting outlet or pipe. Evidently these must vary with the conditions.

As the weight of the air is dependent upon its temperature and the barometric and hygrometric conditions, it is evident that the pressure exerted and the power required by a given fan may vary greatly even at constant speed. Under ordinary conditions, however, changes in the height of the barometer and the humidity of the atmosphere have no appreciable effect upon the pressure and power. But the density varies inversely as the absolute temperature, and, therefore, should enter as a factor even in calculations with reference to air at or about ordinary atmospheric temperatures, and must be taken into account when heated air or gases are handled.

By means of the curves in Fig. 22 are graphically illustrated the theoretical relations between the revolutions at which a fan is run and the volume which is discharged, the pressure which is created, and the horse-power which is required.

In selecting a fan, the facts presented in Fig. 22 should be borne in mind. It appears to be so simple to secure increased volume by running a given fan at higher speed that the influence upon the power required is frequently overlooked. If the necessary amount of power is actually furnished, its expenditure will entail great loss in efficiency as compared with that required to operate a fan properly proportioned to the work.

In the design of a fan-wheel to meet given requirements it is necessary to make its peripheral speed such as to create the desired pressure, and then so proportion its width as to provide for the required air-volume. In practice the actual

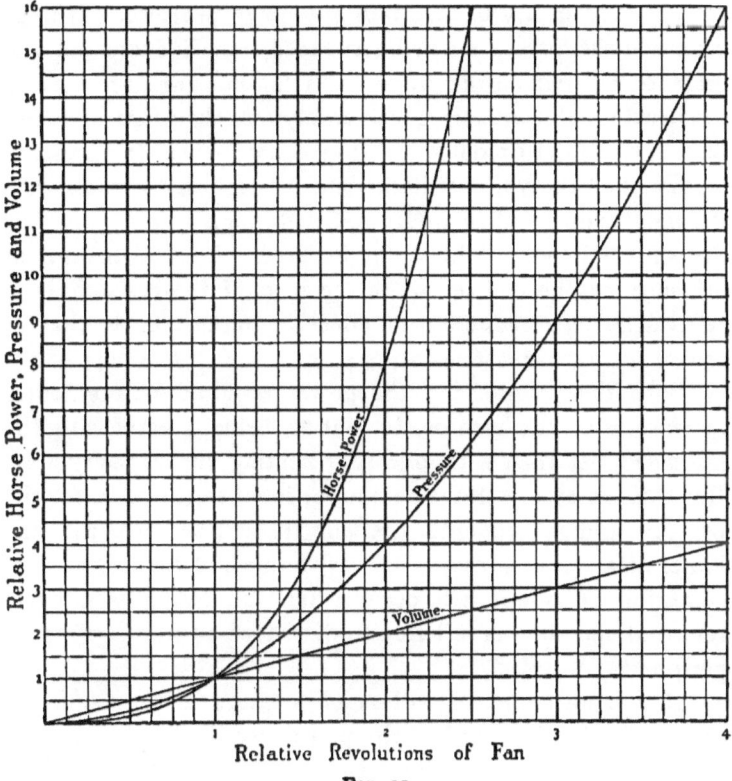

Fig. 22.

amount of air delivered is largely dependent upon the fact of the wheel being encased, the character and dimensions of the case, and the size and resistances of the passages through which the air is conducted. The equivalent of such resistances is in boiler-practice usually represented by the grates, the fuel, tubes, etc., and may evidently be so great at times as to very seriously reduce the theoretical air-discharge of the fan.

It has been determined experimentally that a peripheral-discharge fan, if enclosed in a case, has the ability, if driven to a certain speed, to maintain the pressure corresponding to its tip velocity over an effective area which is usually denominated the "square inches of blast." This area is the limit of its capacity to maintain the given pressure. If it be increased the pressure will be reduced, but if decreased the pressure will remain the same. As fan-housings are usually constructed, this area is considerably less than that of either the regular inlet or outlet. As a consequence, the pressure is lower and the volume discharged is somewhat greater with the regular outlet than would result through an outlet having the square inches of blast for its area. But the maximum pressure may be realized when the sum of the resistances is equivalent to a reduction of effective outlet area to that of square inches of blast.

Both the volume and the power required will evidently increase with the area of the outlet, being greater with the normal outlet than with one representing the capacity area. But this increase will not be proportional to the area, for the pressure and consequently the velocity will be lower with the larger area. The greatest delivery of air and the largest consumption of power will occur when the casing is entirely removed and the fan left free to discharge entirely around its periphery.

If volume alone, regardless of pressure, is the requisite, the larger the fan, the less the power required. If possible, a fan should never be made so small that it is necessary to run it above the required pressure in order to deliver the necessary volume. To double the volume under such circumstances requires eight times the power; three times the volume demands twenty-seven times the power.

When a fan is employed for exhausting hot air or gases, the speed required to maintain a given pressure difference is

MECHANICAL DRAFT.

TABLE No. 109.

RELATION OF VOLUME, WEIGHT, AND PRESSURE OF AIR AND SPEED AND POWER OF A FAN WITH AIR AT DIFFERENT TEMPERATURES.

Temperature, Degrees Fahrenheit.	Volume for Same Weight.	Weight for Same Volume.	Speed Fan to Handle Same Weight.	Pressure-difference due to Speed Necessary to Handle Same Weight.	Pressure-difference for Same Speed and Volume.	Speed to Produce Same Pressure-difference.	Power for Same Speed and Volume.	Power for Speed Necessary to Handle Same Weight.	Power Necessary to Handle Same Weight at Same Pressure with a Properly Proportioned Fan.
1	2	3	4	5	6	7	8	9	10
30	0.96	1.04	0.96	0.96	1.04	0.98	1.04	0.92	0.96
40	0.98	1.02	0.98	0.98	1.02	0.99	1.02	0.96	0.98
50	1.00	1.00	1.00	1.00	1.00	1.00	1.00	1.00	1.00
60	1.02	0.98	1.02	1.02	0.98	1.01	0.98	1.04	1.02
70	1.04	0.96	1.04	1.04	0.96	1.02	0.96	1.08	1.04
80	1.06	0.94	1.06	1.06	0.94	1.03	0.94	1.12	1.06
90	1.08	0.93	1.08	1.08	0.93	1.04	0.93	1.17	1.08
100	1.10	0.91	1.10	1.10	0.91	1.05	0.91	1.21	1.10
125	1.15	0.87	1.15	1.15	0.87	1.07	0.87	1.32	1.15
150	1.20	0.84	1.20	1.20	0.84	1.09	0.84	1.43	1.20
175	1.24	0.81	1.24	1.24	0.81	1.11	0.81	1.55	1.24
200	1.29	0.78	1.29	1.29	0.78	1.14	0.78	1.67	1.29
225	1.34	0.75	1.34	1.34	0.75	1.16	0.75	1.80	1.34
250	1.39	0.72	1.39	1.39	0.72	1.18	0.72	1.93	1.39
275	1.44	0.69	1.44	1.44	0.69	1.20	0.69	2.07	1.44
300	1.49	0.67	1.49	1.49	0.67	1.22	0.67	2.22	1.49
325	1.54	0.65	1.54	1.54	0.65	1.24	0.65	2.36	1.54
350	1.59	0.63	1.59	1.59	0.63	1.26	0.63	2.51	1.59
375	1.63	0.61	1.63	1.63	0.61	1.28	0.61	2.68	1.63
400	1.68	0.59	1.68	1.68	0.59	1.30	0.59	2.84	1.68
425	1.73	0.58	1.73	1.73	0.58	1.32	0.58	3.01	1.73
450	1.78	0.56	1.78	1.78	0.56	1.34	0.56	3.18	1.78
475	1.83	0.55	1.83	1.83	0.55	1.35	0.55	3.35	1.83
500	1.88	0.53	1.88	1.88	0.53	1.37	0.53	3.56	1.88
525	1.93	0.52	1.93	1.93	0.52	1.39	0.52	3.71	1.93
550	1.98	0.51	1.98	1.98	0.51	1.41	0.51	3.92	1.98
575	2.03	0.49	2.03	2.03	0.49	1.43	0.49	4.12	2.03
600	2.08	0.48	2.08	2.08	0.48	1.44	0.48	4.32	2.08
625	2.13	0.47	2.13	2.13	0.47	1.46	0.47	4.54	2.13
650	2.18	0.46	2.18	2.18	0.46	1.48	0.46	4.75	2.18
675	2.22	0.45	2.22	2.22	0.45	1.49	0.45	4.93	2.22
700	2.27	0.44	2.27	2.27	0.44	1.51	0.44	5.16	2.27
725	2.32	0.43	2.32	2.32	0.43	1.52	0.43	5.39	2.32
750	2.37	0.42	2.37	2.37	0.42	1.54	0.42	5.62	2.37
775	2.42	0.41	2.42	2.42	0.41	1.56	0.41	5.86	2.42
800	2.47	0.40	2.47	2.47	0.40	1.57	0.40	6.10	2.47

evidently greater than that necessary when cold air is handled, the difference being due to, and inversely proportional to, the absolute temperature.

As it is the province of a fan to operate at certain velocities in order to produce desired pressures, it is manifest that the temperature of the air, as an influence upon its density, must always be taken into consideration.

Table No. 109 serves to simplify calculation. The temperatures in the table are given in degrees Fahrenheit above zero. These results are also graphically presented in Fig. 23, the curves representing the values in the columns as designated in the table.

The basis upon which the various relations have been calculated is as follows, the atmospheric temperature being taken at 50° F. and absolute zero at 461° below zero F.:

Column 2. The volume of the same weight of air is directly proportional to its absolute temperature.

Column 3. The weight of the same volume of air is inversely proportional to its absolute temperature.

Column 4. The speed of the same fan necessary to handle the same weight of air is directly proportional to its absolute temperature.

Column 5. The pressure-difference due to the speed of the same fan necessary to handle the same weight of air is directly proportional to the square of the speed and inversely proportional to the absolute temperature of the air.

Column 6. The pressure-difference due to the same speed of fan handling the same volume of air is inversely proportional to its absolute temperature.

Column 7. The speed necessary to produce the same pressure-difference is directly proportional to the square root of the absolute temperature.

Column 8. The power required for the same speed and volume is inversely proportional to the absolute temperature.

Column 9. The power required to operate the same fan at the speed necessary to handle the same weight of air is directly proportional to the cube of the speed and inversely proportional to the absolute temperature.

Column 10. The power necessary to handle the same

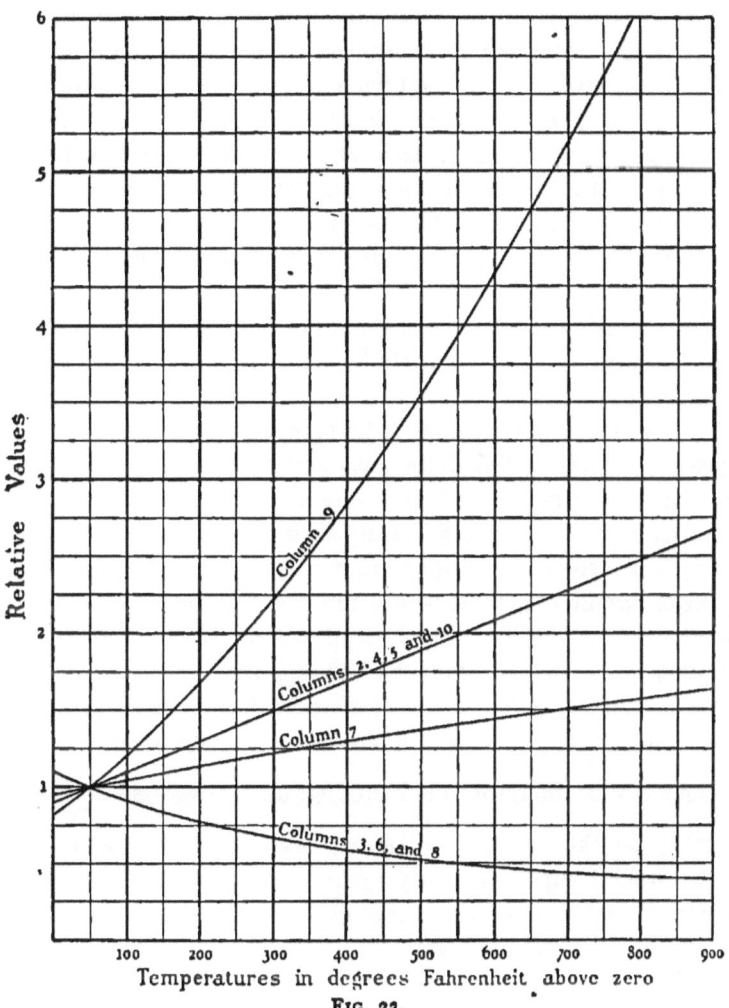

FIG. 23.

weight of air at the same pressure-difference by means of a properly proportioned fan is directly proportional to the speed and to the area required for the passage of the given weight corresponding to the speed. That is, it is directly proportional to the absolute temperature.

The conditions indicated in column 10 are substantially those usually presented in induced-draft practice, wherein it is generally necessary to move the same weight of air under the same pressure, but at a higher temperature than would be required in the case of air handled for forced draft. From column 4 it is evident that if with the same fan it be attempted to handle the same weight at increased temperature, the speed must also be increased; and it is further shown, per column 9, that the power required under these circumstances rises very rapidly with the temperatures. But a fan so designed as to move the same weight without exceeding the pressure created at the lower pressure will require far less power.

Thus suppose that a fan is to be designed to handle air or gases (they both being here considered, for simplicity, as of the same specific gravity) at a temperature of 300°. At this temperature the speed necessary to maintain the same pressure is relatively 1.22, per column 7. But through a unit area of opening, such as would represent the capacity area of a proper fan for the lower temperature, the actual weight moved would be the product of the velocity and the density; that is, only $1.22 \times 0.67 = 0.82$ at 300°. In order to maintain the weight at unity, the area of discharge for the given velocity must therefore be increased until it becomes $\frac{1.0}{0.82} =$ 1.22. So that the power actually expended in moving the same weight with a properly designed fan would for the same pressure be, at 300°, the combined product of the unit pressure, the area, and the velocity; that is, $1 \times 1.22 \times 1.22 =$ 1.49 times that required for a fan properly designed to maintain the same pressure and handle the weight of air at 50°.

If the same diameter of fan be employed under each condition, and the greater volume, in the case of the higher temperature, be provided for by widening the fan, the power required per revolution will be in the relation of 1 at 50° to 1.22 at 300° temperature. For, as already shown, the total power required at the latter temperature is 1.49, while the speed is 1.22, both relatively to the conditions at 50°; hence the power per revolution is $\frac{1.49}{1.22} = 1.22$. From this it is evident that if in each case the fan be driven by a direct-connected engine, whose revolutions of necessity correspond to those of the fan with which it operates in unison, the engine will require additional piston-area, or mean effective pressure, to the extent of only 22 per cent.

From Table No. 109 it is evident that, if a fan be designed to handle a given weight and corresponding volume of air at 50° under a given pressure and with the expenditure of a given amount of power, the following conditions will hold, if the temperature of the air be raised to 300°, for instance. If the same weight is to be handled, the same fan will (per column 4) have to be run at 1.49 times the speed. At this speed the pressure-difference produced will (per column 5) be 1.49 times, and the power expended (per column 9) 2.22 times, that under the first condition. For the same speed and volume the weight handled will (per column 3) be 0.67, the pressure (per column 6) 0.67, and the power (per column 8) 0.67 of that at a temperature of 50°; and to produce the same pressure-difference the fan will have to run (per column 7) at 1.22 times the speed required at 50°. From this it is further evident that in the attempt to handle, with a given fan, the same weight of air at a higher temperature it is necessary to increase the speed above that required to produce the same pressure-difference; and that the power expended is correspondingly and unnecessarily increased.

It is, therefore, obvious that the fan should be designed to meet the specific conditions. Thus a properly proportioned fan will produce the same pressure and handle the same weight of air at 300° with 1.49 times the power (per column 10) required to obtain the same results with the air at 50°; that is, with $\frac{1.49}{2.22} = 0.67 =$ only 67 per cent of that which would have been necessary had the same fan designed for 50° been used in both cases.

These are purely theoretical relations, but they hold substantially under practical conditions. The leakage of air through boiler-settings and the decrease of efficiency through losses in power-transmission, although they affect, yet do not properly enter into, these relations, but must be provided for by additional capacity in fan and engine. Similarly, the increase of volume due to the products of combustion may be provided for.

Methods of Application.—The methods of application of mechanical draft may be broadly classified under two heads, —the plenum and the vacuum methods. Although both were experimented upon by Stevens in 1827 and in the succeeding years, yet the former remained for a long period practically the only form in which mechanical draft was applied. As the term implies, the air under the plenum method is forced through the fire; that is, the pressure maintained below the fire is greater than that of the atmosphere; hence the general term "forced" draft.

Under the plenum method the air may be supplied in either of two ways. First, by making the ash-pit practically air-tight, and then forcing into it the air in sufficient quantity and under the requisite pressure. Evidently, the only escape for the air being through the fuel, it must all be utilized for the purposes of combustion. Second, by making the fire-room itself practically air-tight and maintaining therein the required air-pressure by means of a fan of sufficient capacity to constantly make good the amount of air which under pressure passes to the ash-pits and thence through the fuel.

Under the vacuum method there is practically only one means of application,—that by the introduction of an exhausting-fan in the place of a chimney. This is commonly known as the "induced" or "suction" method. The fan thus serves to maintain the vacuum which would exist if a chimney were employed, and its capacity can be made such as to handle the gases which result from the processes of combustion. A short and comparatively light stack usually serves to carry these gases sufficiently high to permit of their harmless escape to the atmosphere.

Closed Ash-pit System.—This system was naturally first applied because of its ready adaptability to existing conditions. And for the same reason it has, to a great extent, been the system introduced wherever, in an existing plant, it has been desired to burn a cheaper grade of fuel or to add to the steaming capacity. As most simply applied for stationary-boiler work, the air has been introduced in the side of the ash-pit setting through a pipe from the fan. There is a tendency, however, with such a simple arrangement, to fail to properly distribute the air in the ash-pit. The result of unequal distribution, as may occur with improper introduction of the air, is a tendency to blow holes through the fire and overheat the grate-bars wherever the draft is concentrated. Of course, the more intense the draft the greater this tendency. This may readily be overcome by properly deflecting the air by means of an ash-pit damper, so as to insure its thorough distribution throughout the ash-pit before it rises to the grate. Such a damper may be placed either in the bottom of the ash-pit and arranged to receive its air from an underground duct in front of the boilers; or it may be located in the front of the bridge-wall, with the hinge of the door or damper above, so as to deflect the air downward when it is opened. In either case the damper can be readily operated by a rod from the boiler-front. This arrangement necessitates keeping the ash-pit doors closed.

It is evident that for equivalent results it is only necessary for a fan to maintain in the ash-pit an excess of pressure which is equal to the lack of pressure or vacuum which would be produced by a chimney. But in practice the fan generally maintains a greater pressure-difference. The conditions of draft-pressure and the results obtained in connection with certain special devices have already been presented in Tables Nos. 41, 56, and 96.

In marine boilers with internal fire-boxes such arrangements as have just been described are usually inadmissible. There is, however, a method of introduction through the back end of the ash-pit, the air being conducted thereto from the back of the boiler by means of a passage specially provided and extending through the water-back and combustion-chamber. Ordinarily, however, the air is admitted through the ash-pit doors or the openings provided for them. This necessitates a removable arrangement so that the ash-pits may be cleaned.

The pressure within the ash-pit and the furnace-chamber causes all leakage to be outward. The tendency is, therefore, to blow the ashes out of the ash-pit and the flame, smoke, and fuel out of the fire-doors, but with slight effect in the case of stationary boilers at moderate rates of combustion. In the marine service, in order to avoid inconvenience from this source, boilers are frequently fitted with false fronts, within which the air-pressure is maintained. By a proper arrangement of double doors and dampers the disadvantage and danger from flame are completely overcome. The mere arrangement of dampers so connected to the doors that they close when the fire-doors are opened is of great advantage. The false front further presents an excellent yet simple means of admitting air above the fuel, a feature which enters into most arrangements of this character.

Although the closed fire-room system has been far more extensively applied in the naval marine than has the closed ash-pit system, it has been due largely to the existing conditions; and there can be no doubt that of the two the latter arrangement is more satisfactory when the conditions permit of its introduction. The absence of the protective deck, the possibility of open fire-rooms, and the greater space which is usually available generally make possible the closed ash-pit system in the merchant marine, and thereby insure clean and comfortable fire-rooms.

This method of application also presents an opportunity which it shares with the induced system for utilizing the heat of the waste gases, which cannot be attained with the closed fire-room system.

Closed Fire-room System.—This system, in which a plenum condition is maintained in a practically air-tight fire-room, is evidently impracticable for general boiler-practice on land. Its existence is in fact largely due to the exigencies of modern naval warfare.

In the case of the war-ship its maximum steaming capacity is seldom demanded, and then only for a comparatively short time, as during an engagement. Consequently mechanical draft, as applied to vessels of this class, is mainly auxiliary in its character, but nevertheless a practical necessity. By its employment it is possible to construct the machinery within the limits of space and weight which are sufficient for ordinary service, while the reserve of power is stored in the light fans and fittings, instead of in the cumbrous boilers and machinery.

Under ordinary cruising conditions, when the stacks are of moderate height, the fans may not be required, but they must be of form, construction, and capacity sufficient to meet at an instant's notice the maximum demand that may be made upon them. For instance, a vessel of the cruiser type, which may be required in case of necessity to develop 9000 to 10,000 horse-power, may at the usual cruising speed of 10 or 12 knots require only 1500 to 2000 horse-power.

The conditions are well exemplified in Table No. 110,* which presents the results of a series of tests of U. S. S. Charleston at various rates of speed. The rapidly increasing horse-power with increased speed is to be noted, as is also the far more rapid increase in the power of the blowers, made necessary to meet the requirements. The relation between

---

* Transactions Society of Naval Architects and Marine Engineers, 1893.

the actual efficiency, as shown in the coal per indicated horse-power and the knots per ton of coal, is of interest as indicating the difficulties in the way of obtaining even a slight increase of speed when it is well up to the maximum.

TABLE No. 110.

RESULTS OF TESTS OF U. S. S. CHARLESTON.

Items.	Speed in Knots.					
	13	14	15	16	17	18
I. H. P., main engines	2220	2820	3550	4370	5220	6120
Coal per I. H. P. per hour	2.2	2.1	2.0	1.9	2.1	2.5
I. H. P. of blowers	0.0	2.5	6.4	17.6	36.8	69.6
Knots speed per ton of coal	4.34	4.00	3.68	3.41	2.84	2.22

Among the advantages of the closed fire-room system is that of preventing all escape of flame and smoke to the fire-room, the leakage being all inward to the fires. The objection is frequently brought against this system that its use is injurious to the boilers, tending to cause leakage at the tube-plate when the doors are opened. This is generally considered to be due to the chilling action of the comparatively cold air, which thus rushes directly through the furnace-chamber to the tubes. The leakage which sometimes occurs appears to be due to the buckling of the tube-plate, whereby the joints between the tubes and tube-plate are opened up, as a result of the unequal contraction under the chilling effect of the air.

The attempt has been to palliate rather than overcome this difficulty by the introduction of protecting ferrules at the tube-ends. This tendency to leakage under heavy air-pressure should not of necessity be charged against the method of producing the draft. "Because," as stated by an experienced marine-boiler builder,* " a certain boiler does not stand a

---

* " Boiler-construction Suitable for Withstanding the Strains of Forced Draft so far as it Affects the Leakage of Boiler-tubes." A. F. Yarrow. Transactions Institution of Naval Architects. London, 1891.

given air-pressure without the tubes leaking only proves that this air-pressure is too much for that boiler, but does not prove that this air-pressure is too much for every boiler, especially in face of the fact that locomotive-engines are working all over the world at air-pressures varying from 3 inches to 8 inches." There can be no question that the problem, when fairly considered, can be solved so as to avoid any trouble from this source. It can to a great degree be overcome by an arrangement of dampers and fire-doors, such that the draft may be shut off while the doors are open.

Induced System.—This system, whereby a partial vacuum is produced within the furnace, is substantially the same in its effect as a chimney, which it most closely imitates in its action. It possessès, however, not only greater intensity, but other advantages that do not pertain to the chimney. In fact, it is the most natural method of applying mechanical draft, there is no change in arrangement of the boilers from that which obtains when a chimney is used, and there can be no question as to its adaptability when consideration is given to the high pressure-differences and rates of combustion which are secured in a similar manner in locomotive-practice. Its leakage is always inward, avoiding inconvenience from flame and smoke at the fire-doors, which, however, is only liable to occur under heavy air-pressures and when proper damper arrangements are lacking. On shipboard it produces excellent ventilation with open fire-rooms, thereby reducing their temperature. It is cleanly, lends itself readily to control by the dampers which may be introduced for the purpose, and can, like forced draft, be rendered absolutely automatic, requiring no attention whatever from the fireman.

The early objection to this system was that the fans could not stand the high temperature of the gases passing through them. But numbers have now been running for years under these conditions, handling gases from 300° to 500° in temperature, and even higher. Of course these fans require to be of special construction to withstand the heat, and must be provided with means for keeping the bearings cool. If kept dry there is little danger from the action of a sulphurous coal.

The inherent relative efficiencies of the forced and induced systems are difficult of determination because of the many modifying circumstances which exert their influence. Certain comparative tests have, however, been made, principally upon marine boilers; but they can hardly be taken as absolutely conclusive. Among them are those conducted at the Portsmouth (England) Dockyard * in 1890, the results of which are summarized in Table No. 111. Both tests were made on

TABLE No. 111.

RESULTS OF EXPERIMENTS AT PORTSMOUTH DOCKYARD WITH BOILERS OF H. M. S. POLYPHEMUS.

Kind of Draft.	Duration. Hours.	Average Steam-pressure. Pounds.	Temperature.		Total Coal Consumed. Pounds.	Total Water Evaporated. Pounds.	Lbs. Water Evaporated per lb. Coal.		Pounds of Coal Consumed per Hour per Square Foot of Grate.	Lbs. Water Evap. per h. per Sq. Ft. of Grate.		Approximate I. H. P.
			Feed.	Atmosphere.			At Actual Temperature.	At 212°.		At Actual Temperature.	At 212°.	
Induced..	96	74.2	62°	69.90°	80600	777044	9.64	11.13	40.4	389.6	450.4	426.
Forced...	96	77.3	51	49.8	94500	759338	8.03	9.3	47.3	381.	444.	395.

the same boiler, the induced draft being dismantled and replaced by the forced draft. To the decreased tendency of induced draft to blow holes through, as compared with forced draft, and the better regulation and distribution of the air, may undoubtedly be attributed a part of its superiority. It is also claimed that the action of the air and gases upon entering the tube-ends is such that with forced draft a contracted vein is formed which prevents close contact, while with induced draft the partial vacuum in the tubes causes the

---

* Transactions Institution of Naval Architects. London, 1895.

flow to be uniform across the section of the tube, and thereby insures better contact between the gases and the tube surface and readier abstraction of heat.

The induced system presents a ready opportunity for the introduction of air- or water-heaters which are to abstract the heat from the waste gases, thereby securing one of the greatest possible economies in the modern boiler-plant. The economic results actually secured by such an arrangement have already been presented in Table No. 73. The reduction in temperature which thus results not only increases the efficiency of the plant, but has an appreciable effect upon the proportions of the fan; for upon the temperature of the air and gases which pass through the fan must depend its size and speed to accomplish the desired results in the way of draft-production. In a fan operating under the plenum or forced system, the volume of air supplied to the fire is substantially the same as that delivered by the fan, making no allowance for leakage. But with induced draft the fan must handle a volume of air and gases which, although the same in weight, is greater in volume practically in proportion to its increase in absolute temperature. Disregarding leakage, the weight is greater than the air admitted by the portion of the coal which has entered into chemical combination with it. On a basis of 18 pounds of air per pound of coal, this additional weight amounts to 5.5 per cent, while with a supply of 24 pounds it is 4.2 per cent. This increased amount may enter into any refined calculations of fan-capacity, but it is unnecessary to go into the refinement of making allowance for difference in specific gravity or for moisture in the air or fuel.

As the capacity of a fan, both in volume handled and pressure produced, is easily varied by a change in speed, sufficient accuracy is secured in ordinary design by considering that air is the fluid handled, and that the general relations presented in Table No. 109 will exist between volume. weight, and pressure, and the speed and power of a fan at different temperatures.

Advantages of Mechanical Draft.—The simplicity of the chimney is one of its strongest claims to favor as a draft-producer, and it is true that under many conditions it may prove to be, all things considered, the best means of creating the required air-movement. But it is likewise true that under other conditions,—and they are prevalent in modern boiler-practice,—the production of draft by mechanical means may secure results that point to it as the more desirable means. Its influence upon certain factors in boiler economy has been noted here and there in the preceding pages. Its essential advantages may be here enumerated.

As the relative first cost always exercises an important influence upon the introduction of a new device, this feature will be first considered.

In the accompanying curves, Fig. 24, are presented the relative costs of brick chimneys and of equivalent mechanical draft equipments in a number of boiler-plants widely different in character and rated capacity. In certain of these the cost of the existing chimney is known and that of the complete draft-plant was estimated, while in others the cost of the mechanical-draft installation was determined from the contract price and the expense of a chimney to produce equivalent results was calculated. Costs are shown for both single forced and induced engine-driven fans and for duplex engine-driven plants in which either fan may serve as a relay. An apparatus of this latter type is evidently most complete and is necessarily the most expensive. It finds its greatest use where economizers are employed.

An average of the costs for these nine representative plants shows the total expense for installing a forced-draft plant to be only 18.7 per cent, that of a single induced fan and accessories 26.7 per cent, and that of a complete duplex induced-draft plant 42 per cent of that of a chimney. In each case a short steel-plate stack is included.

For a good steam-boiler plant it is fair to assume the following as average fixed charges:

Interest............................ 5 per cent
Depreciation and repairs........... 4½ "
Insurance and taxes................ 1½ "

Total.......................... 11 per cent

Experience has shown that these figures also hold good for a well-designed mechanical-draft apparatus, which has practically the same durability as a steam-boiler with its accessories, and are so accepted here.

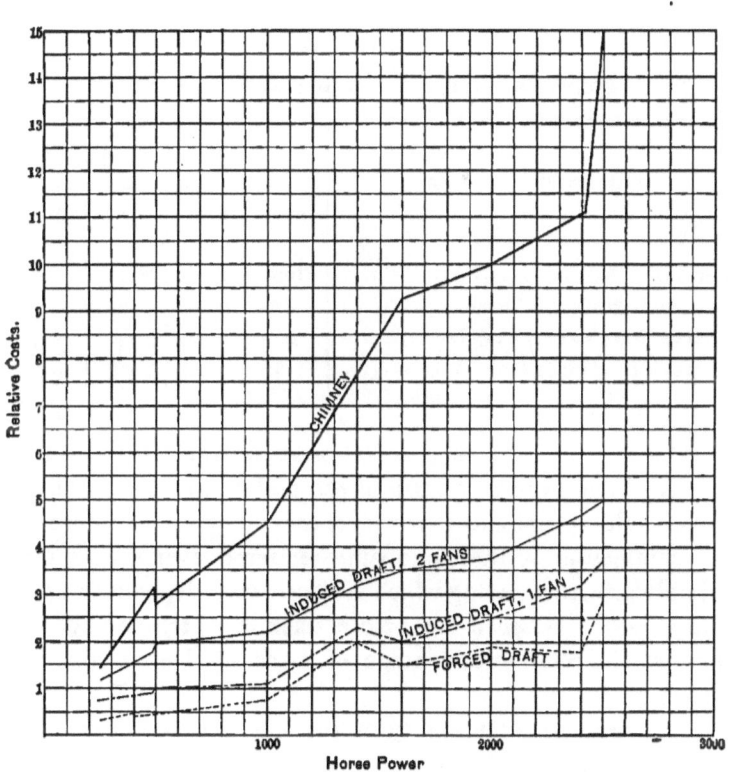

FIG. 24. RELATIVE COSTS OF CHIMNEY DRAFT AND MECHANICAL DRAFT PLANTS.

On the other hand the fixed charges on a chimney may be fairly assumed as:

Interest............................. 5 per cent
Depreciation and repairs............. 1½ "
Insurance and taxes.................. 1½ "

Total........................... 8 per cent

If a brick chimney for a given plant cost $10,000, then on the basis of relative costs and fixed charges just indicated, the relative first cost and annual fixed charges for the different arrangements of mechanical draft would be as presented in Table No. 112.

TABLE No. 112.

COMPARISON OF COSTS AND FIXED CHARGES.

Method of Draft-production.	First Cost.		Annual Fixed Charges.	
	Amount.	Ratio.	Amount.	Ratio.
Chimney..................................	$10000	1.00	$800	1.00
Induced-draft plant (2 fans)..............	4200	0.42	462	0.58
Induced-draft plant (1 fan)..............	2670	0.267	294	0.37
Forced-draft plant (1 fan)..............	1870	0.187	206	0.26

The fact that the mechanical-draft apparatus can usually be placed overhead or on top of the boilers where it occupies no valuable space, and that the space otherwise occupied by the chimney is at the same time rendered available, makes possible a further saving which is necessarily dependent upon land values. When a steel-plate stack is substituted for the chimney the relative advantage of mechanical draft is not so evident.

The peculiar adaptability of the fan and its positive yet flexible action have always been regarded as among its obvious advantages. As it is usually constructed, of steel plate, it may be adapted to any location; and being, in most cases, operated by an independent engine, its action may be rendered independent of climatic conditions, but susceptible to regulation to the exact requirements of the fire.

The intensity of draft produced by a fan makes possible the maintenance of a higher rate of combustion or the burning of low-grade fuels, tends to prevent smoke, and above all renders available for utilization the waste heat of the gases. It may serve as an auxiliary to an overburdened chimney, and under proper conditions may so increase the steaming capacity of the boilers to which it is applied as to make possible a reduction in their number and consequently in the ground-area and first cost for a given output.

APPENDIX.

RULES FOR CONDUCTING BOILER-TRIALS.

THE last report[*] of the Committee of the American Society of Mechanical Engineers, appointed to revise the Code of 1885, whereby standard methods of conducting boiler-trials were established, expresses approval of the conclusions of the former code to the effect that "the standard ' unit of evaporation ' should be one pound of water at 212° Fahr., evaporated into dry steam of the same temperature. This unit is equivalent to 965.7 British thermal units."

It also declares that "the unit of commercial horse-power developed by a boiler shall be taken at $34\frac{1}{2}$ units of evaporation per hour; that is, $34\frac{1}{2}$ pounds of water evaporated per hour from a feed-water temperature of 212° Fahr. into dry steam of the same temperature."

There are here given the general rules for conducting the trial, but the extensive notes of individual members of the Committee regarding the details of methods to be employed are omitted.

PRELIMINARIES TO A TRIAL.

I. *Determine at the outset* the specific object of the proposed trial, whether it be to ascertain the capacity of the

---
[*] Trans. A. S. M. E., vol. xx.

boiler, its efficiency as a steam-generator, its efficiency and its defects under usual working conditions, the economy of some particular kind of fuel, or the effect of changes of design, proportion, or operation; and prepare for the trial accordingly.

II. *Examine the boiler*, both outside and inside; ascertain the dimensions of grates, heating-surfaces, and all important parts; and make a full record, describing the same, and illustrating special features by sketches. The area of heating surface is to be computed from the outside diameter of all tubes, whether water-tubes or fire-tubes. This rule corresponds to the practice of many builders of different types of boilers, and is intended to make the practice of rating heating-surface uniform. All surfaces below the mean water-level which have water on one side and products of combustion on the other are to be considered as water-heating surface, and all surfaces above the mean water-level which have steam on one side and products of combustion on the other are to be considered as superheating surface.

III. *Notice the general condition* of the boiler and its equipment, and record such facts in relation thereto as bear upon the objects in view.

If the object of the trial is to ascertain the maximum economy or capacity of the boiler as a steam-generator, the boiler and all its appurtenances should be put in first-class condition. Clean the heating-surface inside and outside, remove clinkers from the grates and from the sides of the furnace. Remove all dust, soot, and ashes from the chambers, smoke-connections, and flues. Close air-leaks in the masonry and poorly fitted cleaning-doors. See that the damper will open wide and close tight. Test for air-leaks by firing a few shovels of smoky fuel and immediately closing the damper, observing the escape of smoke through the crevices.

IV. *Determine the character of the coal* to be used. For tests of the efficiency or capacity of the boiler the coal should,

APPENDIX.   277

if possible, be of some kind which is commercially regarded as a standard. For New England and that portion of the country east of the Allegheny Mountains, good anthracite egg coal, containing not over 10 per cent of ash, and semi-bituminous Cumberland (Md.) and Pocahontas (Va.) coals are thus regarded. West of the Allegheny Mountains, Pocahontas (Va.) and New River (W. Va.) semi-bituminous, and Youghiogheny or Pittsburg bituminous coals are recognized as standards.* There is no special grade of coal mined in the Western States which is widely recognized as of superior quality or considered as a standard coal for boiler-testing. Big Muddy lump, an Illinois coal mined in Jackson County, Ill., is suggested as being of sufficiently high grade to answer the requirements in districts where it is more conveniently obtainable than the other coals mentioned above.

V. *Establish the correctness of all apparatus* used in the test for weighing and measuring. These are:

1. Scales for weighing coal, ashes, and water.
2. Tanks or water-meters for measuring water. Water-meters, as a rule, should only be used as a check on other measurements. For accurate work, the water should be weighed or measured in a tank.
3. Thermometers and pyrometers for taking temperatures of air, steam, feed-water, waste gases, etc.
4. Pressure-gauges, draft-gauges, etc.

The kind and location of the various pieces of testing apparatus must be left to the judgment of the person conducting the test, always keeping in mind the main object, i.e., to obtain authentic data.

VI. *See that the boiler and chimney are thoroughly heated* before the trial to their usual working temperature. If the

---

* These coals are selected because they are about the only coals which contain the essentials of excellence of quality, adaptability to various kinds of furnaces, grates, boilers, and methods of firing, and wide distribution and general accessibility in the markets.

boiler is new and of a form provided with a brick setting, it should be in regular use at least a week before the trial, so as to dry and heat the walls. If it has been laid off and become cold, it should be worked before the trial until the walls are well heated.

VII. *The boiler and connections* should be proved to be free from leaks before beginning a test, and all water connections, including blow and extra feed-pipes, should be disconnected, stopped with blank flanges, or bled through special openings beyond the valves, except the particular pipe through which water is to be fed to the boiler during the trial. During the test the blow-off and feed-pipes should remain exposed.

If an injector is used, it should receive steam directly through a felted pipe from the boiler being tested.*

See that the steam-main is so arranged that water of condensation cannot run back into the boiler.

VIII. *Starting and Stopping a Test.*—A test should last at least ten hours of continuous running. A longer test may be made when it is desired to ascertain the effect of widely varying conditions, or the performance of a boiler under the working conditions of a prolonged run. The conditions of the boiler and furnace in all respects should be, as nearly as possible, the same at the end as at the beginning of the test. The steam-pressure should be the same; the water-level the same; the fire upon the grates should be the same in quantity and condition; and the walls, flues, etc., should be of the same temperature. Two methods of obtaining the de-

---

* In feeding a boiler undergoing test with an injector taking steam from another boiler, or the main steam-pipe from several boilers, the evaporative results may be modified by a difference in the quality of the steam from such source compared with that supplied by the boiler being tested, and in some cases the connection to the injector may act as a drip for the main steam-pipe. If it is known that the steam from the main pipe is of the same quality as that furnished by the boiler undergoing the test, the steam may be taken from such main pipe.

sired equality of conditions of the fire may be used, viz.: those which were called in the Code of 1885 "the standard method" and "the alternate method," the latter being employed where it is inconvenient to make use of the standard method.

IX. *Standard Method.*—Steam being raised to the working pressure, remove rapidly all the fire from the grate, close the damper, clean the ash-pit, and as quickly as possible start a new fire with weighed wood and coal, noting the time and the water-level while the water is in a quiescent state, just before lighting the fire.

At the end of the test remove the whole fire, which has been burned low, clean the grates and ash-pit, and note the water-level when the water is in a quiescent state, and record the time of hauling the fire. The water-level should be as nearly as possible the same as at the beginning of the test. If it is not the same, a correction should be made by computation, and not by operating the pump after the test is completed.

X. *Alternate Method.* — The boiler being thoroughly heated by a preliminary run, the fires are to be burned low and well cleaned. Note the amount of coal left on the grate as nearly as it can be estimated; note the pressure of steam and the water-level, and note this time as the time of starting the test. Fresh coal which has been weighed should now be fired. The ash-pits should be thoroughly cleaned at once after starting. Before the end of the test the fires should be burned low, just as before the start, and the fires cleaned in such a manner as to leave the bed of coal of the same depth, and in the same condition, on the grates as at the start. The water-level and steam-pressures should previously be brought as nearly as possible to the same point as at the start, and the time of ending of the test should be noted just before fresh coal is fired. If the water-level is not the same as at the start, a correction should be made by computation, and not by operating the pump after the test is completed.

XI. *Uniformity of Conditions.*—In all standard trials the conditions should be maintained uniformly constant. Arrangements should be made to dispose of the steam so that the rate of evaporation may be kept the same from beginning to end. This may be accomplished in a single boiler by carrying the steam through a waste steam-pipe, the discharge from which can be regulated as desired. In a battery of boilers in which only one is tested the draught can be regulated on the remaining boilers, leaving the test-boiler to work under a constant rate of production.

Uniformity of conditions should prevail as to the pressure of steam, the height of water, the rate of evaporation, the thickness of fire, the times of firing and quantity of coal fired at one time, and as to the intervals between the times of cleaning the fires.

XII. *Keeping the Records.*—Take note of every event connected with the progress of the trial, however unimportant it may appear. Record the time of every occurrence and the time of taking every weight and every observation.

The coal should be weighed and delivered to the fireman in equal proportions, each sufficient for not more than one hour's run, and a fresh portion should not be delivered until the previous one has all been fired. The time required to consume each portion should be noted, the time being recorded at the instant of firing the last of each portion. It is desirable that at the same time the amount of water fed into the boiler should be accurately noted and recorded, including the height of the water in the boiler, and the average pressure of steam and temperature of feed during the time. By thus recording the amount of water evaporated by successive portions of coal, the test may be divided into several periods if desired, and the degree of uniformity of combustion, evaporation, and economy analyzed for each period. In addition to these records of the coal and the feed-water, half-hourly observations should be made of the temperature of the feed-water, of the flue gases, of the external air in the boiler-room, of the temperature of the furnace when a furnace-pyrometer is used, also of the pressure of steam, and of the readings of the instruments for determining the moisture in the steam. A log should be kept on properly prepared blanks containing columns for record of the various observations.

When the "standard method" of starting and stopping the test is used, the hourly rate of combustion and of evaporation and the horse-power may be computed from the records taken during the time when the fires are in active condition. This time is somewhat less than the actual time which elapses between the beginning and end of the run. This method of computation is necessary, owing to the loss of time due to kindling the fire at the beginning and burning it out at the end.

XIII. *Quality of Steam.*—The percentage of moisture in the steam should be determined by the use of either a throttling or a separating steam-calorimeter. The sampling-nozzle should be placed in the vertical steam-pipe rising from the boiler. It should be made of $\frac{1}{2}$-inch pipe, and should extend across the diameter of the steam-pipe to within half an inch of the opposite side, being closed at the end and perforated with not less than twenty $\frac{1}{8}$-inch holes equally distributed along and around its cylindrical surface, but none of these holes should be nearer than $\frac{1}{4}$ inch to the inner side of the steam-pipe. The calorimeter and the pipe leading to it should be well covered with felting. Whenever the indications of the throttling or separating calorimeter show that the percentage of moisture is irregular, or occasionally in excess of three per cent, the results should be checked by a steam-separator placed in the steam-pipe as close to the boiler as convenient, with a calorimeter in the steam-pipe just beyond the outlet from the separator. The drip from the separator should be caught and weighed, and the percentage of moisture computed therefrom added to that shown by the calorimeter.

Superheating should be determined by means of a thermometer placed in a mercury-well or oil-well inserted in the steam-pipe.

For calculations relating to quality of steam and corrections for quality of steam. (See page 119.)

XIV. *Sampling the Coal and Determining its Moisture.*— As each barrow-load or fresh portion of coal is taken from the coal-pile, a representative shovelful is selected from it and placed in a barrel or box in a cool place and kept until the end of the trial. The samples are then mixed and broken into pieces not exceeding one inch in diameter, and reduced by the process of repeated quartering and crushing until a final sample weighing about five pounds is obtained, and the size of the larger pieces are such that they will pass through a sieve with $\frac{1}{4}$-inch meshes. From this sample two one-quart, air-tight glass preserving-jars, or other air-tight vessels which will prevent the escape of moisture from the sample, are to be promptly filled, and these samples are to be kept for subsequent determinations of moisture and of heating value, and for chemical analyses. During the process of quartering, when the sample has been reduced to about 100 pounds, a quarter to a half of it may be taken for an approximate determination of moisture. This may be made by placing it in a shallow iron pan, not over three inches deep, carefully weighing it, and setting the pan in the hottest place that can be found on the brickwork of the boiler setting or flues, keeping it there for at least twelve hours, and then weighing it. The determination of moisture thus made is believed to be approximately accurate for anthracite and semi-bituminous coals, and also for Pittsburg or Youghiogheny coal; but it cannot be relied upon for coals mined west of Pittsburg, or for other coals containing inherent moisture. For these latter coals it is important that a more accurate method be adopted. The method recommended by the Committee for all accurate tests, whatever the character of the coal, is described as follows:

Take one of the samples contained in the glass jars, crush the whole of it by running it through an ordinary coffee-mill adjusted so as to produce somewhat coarse grains (less than $\frac{1}{16}$ inch), thoroughly mix the crushed sample, select from it a portion of from 10 to 50 grams, weigh it in a balance which will easily show a variation as small as 1 part in 1000, and dry it in an air or sand bath at a temperature between 240 and 280 degrees Fahr. for one hour. Weigh it and record the loss, then heat and weigh it again repeatedly, at intervals of an hour or less, until the minimum weight has been reached and the weight begins to increase by oxidation of a portion of the coal. The difference between the original and the minimum weight is taken as the moisture. This moisture should preferably be made on duplicate samples, and the results should agree within 0.3 to 0.4 of one per cent, the mean of the two determinations being taken as the correct result.

If the coal contains an appreciable amount of surface moisture, another portion of the 100 pounds sample should be weighed and spread out in a thin layer on a clean sheet-iron plate, and exposed for a period of twenty-four hours to the atmosphere of the boiler-room, and by this means air-dried. After being weighed again, the percentage which the weight shrinks during this drying may be termed the percentage of surface moisture.

XV. *Treatment of Ashes and Refuse.*—The ashes and refuse are to be weighed in a dry state. For elaborate trials a sample of the same should be procured for analysis. When it is desired to know accurately the amount of coal consumed, as distinguished from combustible, all lumps of unconsumed coal one-half inch or more in diameter are to be picked from the refuse and deducted from the weight of coal fired.

XVI. *Calorific Tests and Analysis of Coal.*—The quality of the fuel should be determined either by heat test or by analysis, or by both.

The rational method of determining the total heat of combustion is to burn the sample of coal in an atmosphere of oxygen-gas, the coal to be sampled as directed in Article XIV of this Code.

The chemical analysis of the coal should be made only by an expert chemist. The total heat of combustion computed from the results of the ultimate analysis should be obtained by the use of Dulong's formula (with constants modified by recent determinations), viz.,

$$14600\ C + 62000\left(H - \frac{O}{8}\right),$$

in which C, H, and O refer to the proportion of carbon, hydrogen, and oxygen respectively, and determined by the ultimate analysis.*

It is recommended that the analysis and the heat test be each made by two independent laboratories, and the mean of the two results, if there is any difference, be adopted as the correct figures.

It is desirable that a proximate analysis should also be made to determine the relative proportions of volatile matter and fixed carbon in the coal.

XVII. *Analysis of Flue-gases.*—The analysis of the flue-gases is an especially valuable method of determining the relative value of different methods of firing, or of different kinds of furnaces. In making these analyses great care should

---

* Favre and Silbermann give 14544 B. T. U. per pound carbon; Berthelot 14647 B. T. U. Favre and Silbermann give 62032 B. T. U. per pound hydrogen; Thomson, 61816 B. T. U.

be taken to procure average samples, since the composition is apt to vary at different points of the flue; and where complete determinations are desired, the analysis should be intrusted to an expert chemist. For approximate determinations the Orsat* or the Hempel† apparatus may be used by the engineer.

XVIII. *Smoke Observations.*—It is desirable to have a uniform system of determining and recording the quantity of smoke produced where bituminous coal is used. The system commonly employed is to express the degree of smokiness by means of percentages dependent upon the judgment of the observer. The Committee does not place much value upon a percentage method, because it depends so largely upon the personal element, but if this method is used, it is desirable that, so far as possible, a definition be given in explicit terms as to the basis and method employed in arriving at the percentage.

XIX. *Miscellaneous.*—In tests for purposes of scientific research, in which the determination of all the variables entering into the test is desired, certain observations should be made which are in general unnecessary for ordinary tests. These are the measurement of the air-supply, the determination of its contained moisture, the determination of the amount of heat lost by radiation, of the amount of infiltration of air through the setting, and (by condensation of all the steam made by the boiler) of the total heat imparted to the water.

As these determinations are not likely to be undertaken except by engineers of high scientific attainments, it is not deemed advisable to give directions for making them.

XX. *Calculations of Efficiency.*—Two methods of defining

---

* See R. S. Hale's paper on " Flue Gas Analysis," *Transactions A. S. M. E.*, vol. XVIII. p. 901.
† See Hempel on " Gas Analysis."

and calculating the efficiency of a boiler are recommended. They are:

1. Efficiency of the boiler $= \dfrac{\text{Heat absorbed per lb. combustible}}{\text{Heating value of 1 lb. combustible}}$

2. Efficiency of the boiler and grate

$= \dfrac{\text{Heat absorbed per lb. coal}}{\text{Heating value of 1 lb. coal}}$

The first of these is sometimes called the efficiency based on combustible, and the second the efficiency based on coal. The first is recommended as a standard of comparison for all tests, and this is the one which is understood to be referred to when the word "efficiency" alone is used without qualification. The second, however, should be included in a report of a test, together with the first, whenever the object of the test is to determine the efficiency of the boiler and furnace together with the grate (or mechanical stoker), or to compare different furnaces, grates, fuels, or methods of firing.

The heat absorbed per pound of combustible (or per pound coal) is to be calculated by multiplying the equivalent evaporation from and at 212 degrees per pound combustible (or coal) by 965.7.

In calculating the efficiency where the coal contains an appreciable amount of surface moisture, allowance is to be made for the heat lost in evaporating this moisture by adding to the heat absorbed by the boiler the heat of evaporation thus lost. The percentage of surface moisture used in this calculation is that which is found in the manner described in Article XIV of Code.

XXI. *The Heat-balance.*—An approximate "heat-balance," or statement of the distribution of the heating value of the coal among the several items of heat utilized and heat lost may be included in the report of a test when analyses of the fuel and of the chimney gases have been made. It should be reported in the following form:

APPENDIX. 287

HEAT BALANCE, OR DISTRIBUTION OF THE HEATING VALUE OF THE COMBUSTIBLE.

Total Heat Value of 1 lb. of Combustible.................. B. T. U.

	B. T. U.	Per Cent.
1. Heat absorbed by the boiler = evaporation from and at 212 degrees per pound of combustible $\times$ 965.7.		
2. Loss due to moisture in coal = per cent of moisture referred to combustible $\div$ 100 $\times$ [(212 $-$ t) + 966 + 0.48(T $-$ 212)](t = temperature of air in the boiler-room, T = that of the flue gases).		
3. Loss due to moisture formed by the burning of hydrogen = per cent of hydrogen to combustible $\div$ 100 $\times$ 9 $\times$ [(212 $-$ t) + 966 + 0.48(T $-$ 212)].		
4.* Loss due to heat carried away in the dry chimney gases = weight of gas per pound of combustible $\times$ 0.24 $\times$ ($T - t$).		
5.† Loss due to incomplete combustion of carbon = $\dfrac{CO}{CO_2 + CO}$ $\times$ $\dfrac{\text{per cent C in combustible}}{100}$ $\times$ 10150.		
6. Loss due to unconsumed hydrogen and hydrocarbons, to heating the moisture in the air, to radiation, and unaccounted for.		
Totals.................		100.00

* The weight of gas per pound of carbon burned may be calculated from the gas analyses as follows:

Dry gas per pound carbon = $\dfrac{11\,CO_2 + 8\,O + 7(CO + N)}{3(CO_2 + CO)}$, in which CO_2, CO, O, and N are the percentages by volume of the several gases. As the sampling and analyses of the gases in the present state of the art are liable to considerable errors, the result of this calculation is usually only an approximate one. The heat-balance itself is also only approximate for this reason, as well as for the fact that it is not possible to determine accurately the percentage of unburned hydrogen or hydrocarbons in the flue gases.

The weight of dry gas per pound of combustible is found by multiplying the dry gas per pound of carbon by the percentage of carbon in the combustible, and dividing by 100.

† CO_2 and CO are respectively the percentage by volume of carbonic acid and carbonic oxide in the flue gases. The quantity 10150 = No, heat-units generated by burning to carbonic acid one pound of carbon contained in carbonic oxide.

XXII. *Report of the Trial.*—The data and results should be reported in the manner given in the following table, omitting lines where the tests have not been made as elaborately as provided for in such table. Additional lines may be added for data relating to the specific object of the test. The extra lines should be classified under the headings provided in the

288   APPENDIX.

table, and numbered, as per preceding line, with sub letters, *a*, *b*, etc.

DATA AND RESULTS OF EVAPORATIVE TRIALS.

Made by................of..................boiler at....................to determine...
....
Principal conditions governing the trial ....
....
....
Kind of fuel...
State of the weather..
 1. Date of trial ....
 2. Duration of trial... hours.

Dimensions and Proportions.

(A complete description of the boiler should be given on an annexed sheet.)
 3. Grate surface........width.......length.......area...... sq. ft.
 4. Water-heating surface...................................... "
 5. Superheating surface.. "
 6. Ratio of water heating surface to grate surface...........
 7. Ratio of minimum draft area to grate surface..............

Average Pressures.

 8. Steam-pressure by gauge...................................... lbs.
 9. Atmospheric pressure by barometer......................... in.
 10. Force of draft between damper and boiler................. "
 11. Force of draft in furnace.................................... "
 12. Force of draft in ash-pit..................................... "

Average Temperatures.

 13. Of external air... deg.
 14. Of fire room... "
 15. Of steam... "
 16. Of feed water entering heater.............................. "
 17. Of feed water entering economizer........................ "
 18. Of feed water entering boiler............................... "
 19. Of escaping gases from boiler .... "
 20. Of escaping gases from economizer........................ "

APPENDIX. 289

Fuel.

21. Size and condition...
22. Weight of wood used in lighting fire.................... lbs.
23. Weight of coal as fired *................................... "
24. Percentage of moisture in coal †........................ per cent.
25. Total weight of dry coal consumed (Art. XIV, Code)...... lbs.
26. Total ash and refuse...................................... "
27. Total combustible consumed............................... "
28. Percentage of ash and refuse in dry coal................ per cent.

Proximate Analysis of Coal.

                                        Of Coal. Of Combustible.
29. Fixed carbon.............................. per cent. per cent.
30. Volatile matter........................... "   "
31. Moisture.................................. "   ———
32. Ash....................................... "   ———
                                        100 per cent. 100 per cent.
33. Sulphur, separately determined........ "   "

Ultimate Analysis of Dry Coal.
(Art. XVI, Code.)

34. Carbon (C)... per cent.
35. Hydrogen (H).. "
36. Oxygen (O).. "
37. Nitrogen (N).. "
38. Sulphur (S)... "
                                                             100 per cent.
39. Moisture in sample of coal as received..................... "

Analysis of Ash and Refuse.

40. Carbon.. per cent.
41. Earthy matter... "

Fuel per Hour.

42. Dry coal consumed per hour................................ lbs.
43. Combustible consumed per hour............................ "
44. Dry coal per square foot of grate surface per hour........ "
45. Combustible per square foot of water heating surface per hour.. "

---

* Including equivalent of wood used in lighting the fire, not including unburnt coal withdrawn from furnace at end of test. One pound of wood is taken to be equal to 0.4 pound of coal.

† This is the total moisture in the coal as found by drying it artificially, as described in Art. XIV of Code.

APPENDIX.

Calorific Value of Fuel.

46. Calorific value by oxygen calorimeter, per pound of dry coal .... B. T. U.
47. Calorific value by oxygen calorimeter, per pound of combustible .... " " "
48. Calorific value by analysis, per lb. of dry coal* .... " " "
49. Calorific value by analysis, per pound of combustible ..... " " "

Quality of Steam.

50. Percentage of moisture in steam .... per cent.
51. Number of degrees of superheating .... deg.
52. Quality of steam (dry steam = unity) ....
53. Factor of correction for quality of steam (page 119) ....

Water.

54. Total weight of water fed to boiler .... lbs.
55. Water actually evaporated, corrected for quality of steam   "
56. Equivalent water evaporated into dry steam from and at 212 degrees ....   "

Water per Hour.

57. Water evaporated per hour, corrected for quality of steam   "
58. Equivalent evaporation per hour from and at 212 degrees.   "
59. Equivalent evaporation per hour from and at 212 degrees per square foot of water-heating surface ....   "

Horse-power.

60. Horse-power developed. (34¼ lbs. of water evaporated per hour into dry steam from and at 212 degrees, equals one horse-power)† .... H. P.
61. Builders' rated horse power ....   "
62. Percentage of builders' rated horse-power developed ...... per cent.

Economic Results.

63. Water apparently evaporated per lb. of coal under actual conditions. (Item 54 ÷ Item 23) .... lbs.
64. Equivalent evaporation from and at 212 degrees per lb. of coal (including moisture) ....   "
65. Equivalent evaporation from and at 212 degrees per lb. of dry coal ....   "
66. Equivalent evaporation from and at 212 degrees per lb. of combustible ....   "

---

* See formula for calorific value under Article XVI of Code.
† Held to be the equivalent of 30 lbs. of water per hour evaporated from 100 degrees Fahr. into dry steam at 70 lbs. gauge-pressure (See Introduction to Code.)

Efficiency.

(See Art. XX, Code.)

67. Efficiency of the boiler; heat absorbed by the boiler per lb. of combustible divided by the heat-value of one lb. of combustible.*... per cent.
68. Efficiency of boiler, including the grate; heat absorbed by the boiler, per lb. of dry coal fired, divided by the heat value of one lb. of dry coal.†..........................

Cost of Evaporation.

69. Cost of coal per ton of 2240 lbs. delivered in boiler-room... $
70. Cost of fuel for evaporating 1000 lbs. of water under observed conditions....................................... $
71. Cost of fuel used for evaporating 1,000 lbs. of water from and at 212 degrees...................................... $

Smoke Observations.

72. Percentage of smoke as observed.................................
73. Weight of soot per hour obtained from smoke-meter...................
74. Volume of soot obtained from smoke-meter per hour..................

---

* In all cases where the word "combustible" is used, it means the coal without moisture and ash, but including all other constituents. It is the same as what is called in Europe "coal dry and free from ash."

† The heat value of the coal is to be determined either by an oxygen calorimeter or by calculation from ultimate analysis. When both methods are used the mean value is to be taken. (See page 10.)

INDEX.

ABSOLUTE ZERO, 12
Abstractors, 168, 171, 175
Air, 25
  admission above fire, 113
  composition, 25
  efflux, 198
  excess, loss on account of, 37, 115, 119
  expansion by heat, 13
  for dilution, 35, 37, 118, 189
  heater, 168, 175
    tests, 170
  leakage, 225
  measurement, 47
  preheating, 115, 137, 168, 175
  pressure, 195, 200, 263
    and rate of combustion, 218, 224
  properties, 13, 25
  saturated mixtures, 13
  supply, admission above fire, 113
    calculation from gas-analysis, 41
    effect on ideal temperature, 56, 118
    effect on efficiency of fuel, 119, 124, 163
    effect on temperature of gases, 116, 163
    for combustion, 32, 39
    for dilution, 35, 37, 118, 189
    for high rates of combustion, 186
    heating, 115, 137, 168, 175
    insufficiency, influence of, 39
    loss of efficiency due to excess, 119, 121, 163
    measurement, 47
    preheating, 115, 137, 168, 175
    with different rates of combustion, 188
  temperature, influence on movement, 257
  velocity of flow, 195, 200, 203

Air, volume, 13
  weight, 13
  work to move, 201
American Line, test, 171
Anemometer, 47, 207
Anthracite coal, see Coal, anthracite
Artificial fuels, 84
Ash, influence of, 100
A. S. M. E. boiler-trial code, 274
  rating of boilers, 145
Atomic theory, 25

BAGASSE, 60
  calorific value, 61
  composition, 63
  diffusion, 60
  fuel value, 63
  mill, 60
  moisture in, 61
Bituminous coal, see Coal, bituminous
Blowers, see Fans
Boilers, steam
  combustion rate, effect on evaporation, 160, 183
  convection of furnace-heat, 149
  depreciation, 4
  draft, 194-273
    effect of, 132, 156, 270
  economy of high rates of combustion, 183
  efficiency, 1, 140-179
    measure of, 140
    quantitative, 5
    relative, 141, 143, 149
    sources, 155
    ultimate, 3
  evaporation in different sections, 152
    per horse-power, 147
    per square foot of heating-surface, 148, 182
  evaporative performance, 160

INDEX.

Boilers, grate-area, 157
  effect of, 192
  vs. heating-surface, 181
  heat-balance, 125, 142
    disposition of, 154
  heating-surface, 158
    per horse-power, 148
    vs. grate-surface, 181
  horse-power, 149
  leakage of air, 225
  loss due to temperature of gases, 162
  operating expenses, 6
  output, effect of quality of fuel, 127
  powdered-fuel furnaces, 179
  radiation of furnace-heat, 149
  rate of combustion, 180, 216
  rating, 143, 149
  retarders, 172, 224
  saving by burning cheap fuels, 131
  Serve-tubes, 174
  settings, leakage through, 225
    radiation through, 154
    temperature in, 154
  surface-ratio, 149, 191
    and rate of combustion, 191
    influence of, 160
  temperature of fire, 151
    in tubes, 153
  testing, rules for, 274
  tube-heating efficiency, 172
  tubes, ribbed, 175

CALORIE, 15
Calorimeter, fuel, 49
Carbon, 24
  combustion of, 27, 37, 48
  heat of combustion, 48
  ideal temperature, 52
  in ash, 101
  properties, 33
  union with oxygen, 27, 37, 48
Carbonic acid, 28, 37
  properties, 29
Carbonic dioxide, 28
Carbonic oxide, 28
  heat of combustion, 49
  loss of efficiency on account of, 110
  produced by excessive firing, 112
  properties, 29
Centennial boiler rating, 145
Charcoal, 84
Chimney, cost, 242
  cost vs. fan, 271
  draft, see Draft, chimney

Chimney vs. mechanical draft, 244
  efficiency, 243
  vs. fan, 245
  formulæ, 232
  Gale's formulæ, 235
  height for given rate of combustion, 240
  Kent's formula, 234
  Kent's table, 236
  temperature, influence of, 230
  theory, 228
  vs. fan, 244
Chimneys, 228
  capacities, 236
Closed ash-pit system, 263
Closed fire-room system, 265
Coal, see also Fuels
  analysis, 69
  anthracite, 30, 71
    composition, 30, 52, 72, 74
    efficiency, relative, 106, 130
    of small sizes, 106
    rate of combustion, 106, 218
  bituminous, 30, 69
    composition, 30, 69
  caking, 70
  calorific value, 52, 74
  cannel, 70
  classification, size, 67
    geographical, 73
  combustion, see also Combustion
    losses incident to, 126
    of small sizes, 218
  composition, 30, 69, 74-83
    proximate, 69
    ultimate, 69
  cost for equivalent evaporation, 131
    influence of, 128
  economy of cheap coal, 131
  efficiency, 99, 184
    commercial, 127
    comparative, 127, 130
    influences affecting, 122
    of various coals, 130
  evaporation for equivalent cost, 133
  fire, temperature, 151
  formation, 65, 67
  heat of combustion, 48
  lignite, 30, 68
    composition, 30, 52, 68
  losses incident to burning, 129
  moisture in, influence of, 108
  non-caking, 70
  quality, effect of, 127
  rate of combustion, 180, 188, 216
  semi-anthracite, 71

INDEX.

Coal, semi-bituminous, 30, 71
  composition, 71
  size, influence of, 101
  sizes, 103
  small, requisites for burning, 102
  small sizes, combustion of, 218
Coke, 87
Combustion, 24–56
  admission of air above fire, 113
  air required for, 32
  air-supply, effect of, 37, 117, 118, 124, 163, 190
  carbon, 27, 37, 48
  data, 33
  definition, 24
  draft for given rate, 186, 216
  economy of high rates, 183
  effect on evaporation, 160
  efficiency, 184
  forced, *see* Draft, mechanical
  heat of, 48
    calculation, 50
    distribution, 151
    losses, 126, 155
  height of chimney for different rates, 240
  ideal temperature, 52, 117
  influence of ash, 100
    of excessive firing, 112
    of frequent firing, 105
    of moisture in coal, 108
    of size of coal, 101, 106, 218
  losses incident to, 97, 129
    on account of carbonic oxide, 110
    on account of excess of air, 37, 115, 119
    on account of smoke, 108
  products of, 125, 129
  rate of, 180–193, 216
  rate, chimney for, 240
    relatively to draft, 216, 220
    with small anthracite coals, 218
  rates for different boilers, 181
  small coal, 106, 218
  substances concerned, 53
  temperature of, 151
    escaping gases, 121
  thick fire, 193
Cost, primary and fixed charges, 4, 271
Coxe stoker, 217

Dilution, air for, 35, 37
Draft, 194–227
  actual conditions, 211
  and rate of combustion, 218
  artificial, *see* Draft, mechanical
  chimney, 228–246

Draft, Gale's formulæ, 235
  height for given rate of combustion, 240
  Kent's formula, 234
  principles, 228
  temperature, influence on, 230
  theory, 228
  conditions, 210
  definition, 194
  forced, see Draft, mechanical
  gauge, 205
  measurement, 203
  mechanical, 247–273
    advantages, 27
    air-pressure produced, 218
    air-supply, 38
    application for burning small coals, 103, 218
    burning small fuels, 133, 218
    carbonic oxide, prevention of, 111
    closed ash-pit, 263
    closed fire-room, 265
    cost *vs.* chimney, 271
    definition, 247
    economy with economizer, 167
    economy with small coals, 134
    forced, 262
    forced *vs.* induced, 268
    for small coals, 103, 106, 218
    for high rates of combustion, 180
    induced, 267
    induced *vs.* forced, 268
    influence of, 132, 270
    influence on smoke-prevention, 136
    methods, 262
    record, 213
    retarders, influence on, 174
    steam-jets, 247
    vs. chimney-draft, 134, 244
  pressure, 195, 204
  record, 213
  recorder, 208
  regulation, 213
  velocity, 195
  work to move air, 201

Eaves helical-draft boiler, 175
Economizers, 162
  effect of state of surface, 168
  saving by, 165
  tests, 166
Elementary substances, 28
  atomic weights, 28
  symbols, 28
Ellis and Eaves system, 171
  test, 172

Engines, steam, steam per horse-power, 144
Evaporation, factors of, 96
  for equivalent cost of coal, 131
  per horse-power, 147
  per square foot of heating surface, 182
  rate for equivalent cost of coal, 133
  unit of, 94
Excess of air, loss on account of, 37, 115, 119
Exhausters, see Fans

FACTORS of evaporation, 96
Fans, see also Draft, mechanical
  cost vs. chimney, 271
  design, 251
  effect of temperature on results, 257
  power required, 213
  pressure created, 255
  types, 250
  volume moved, 245
Feed-water heaters, 162
Fire, temperature of, 151
  thickness of, 193
Firing, excessive, effect of, 112
  frequency, influence of, 105
  mechanical vs. hand, 178
Fixed charges, 4, 271
Flue-gas, see Gas, flue
Forced draft, see Draft, mechanical
Fuel-gas, 40, 88, 90, 111
Fuels, 57–91
  see also Coal
  air required for combustion, 32
  artificial, 57, 84
  combustion, 29
  composition, 30, 57
  cost, influence of, 128
  definition, 57
  draft required, 225
  economy of cheap, 131
  efficiency, 92–139, 184
    influences affecting, 122
  heat of combustion, 52
  natural, 57
  patent, 90
  powdered, 179
  quality, effect of, 127
Furnace, down-draft, 138
  heat, convection, 149
    radiation, 149
  powdered fuel, 179
  types, 97

GAS-FLUE, ANALYSIS, 40, 111
  composition, 44

Gas-flue, loss due to temperature, 119, 162
  temperature, 153
    with retarders, 173
  tests by Donkin & Kennedy, 38
Gas, fuel, 88
  natural, 84
  producer, 90
  water, 90
Gauge, water, 205
Grate, see Boilers, steam, grate

HEAT-ABSTRACTORS, 168, 171, 175
  balance, 125, 142
  combustion, of, 48
  disposition of, in steam-boilers, 154
  distribution in steam-boilers, 151
  latent, 17
  losses incident to combustion, 126
  mechanical equivalent, 15
  sensible, 17
  specific, 14
  transmission, 150
  unit of, 15
Heating feed-water, saving by, 165
  surface, see Boilers, steam, heating-surface
Hook draft-gauge, 206
Horse-power of boilers, 143, 149
Hydrocarbons, 29, 31
Hydrogen, 26
  heat of combustion, 49
  ideal temperature of combustion, 53
  properties, 33

IDEAL TEMPERATURE of combustion, 52
Induced draft, see Draft, mechanical
Iroquois Hotel, test, 134

LATENT HEAT, 17
Leakage of air, 225
Lignite, see Coal, lignite

MAHLER'S BOMB-CALORIMETER, 48
Marsh-gas, 29
  heat of combustion, 49
Mechanical draft, see Draft, mechanical
Mechanical equivalent of heat, 15
Megass, see Bagasse
Moisture in coal, 108

NATURAL DRAFT, see Draft, chimney
Nitrogen, 25, 37
Natural gas, 90

OLEFIANT GAS, 49

INDEX.

Operating expenses, 6
Orsat apparatus, 40
Oxygen, 24, 37
  union with carbon, 27

PACIFIC MILLS, test, 169
Patent fuel, 90
Peat, 64
  composition, 66
Petroleum, 30, 73, 85
  composition, 30, 85
Pitot's tube, 209
Pressure gauge, 205
  head, 198
  relation to velocity, 195
Producer gas, 90
Proximate analysis, 69

RADIATION OF FURNACE-HEAT, 149
Rate of combustion, 180–193
Retarders, 174, 224

SEMI-ANTHRACITE COAL, see Coal, semi-anthracite
Semi-bituminous coal, see Coal, semi-bituminous
Sensible heat, 17
Serve-tube, 174
Smoke analysis, 109
  loss on account of, 108
  prevention, 108, 135
Specific heat, 14
Steam, 8–18
  boilers, see Boilers, steam
  bulk, 19
  composition, 8
  expansion by heat, 19
  generation, requisites for, 2
  jets, 247
  latent heat, 17, 19
  per horse-power for steam-engines, 144
  properties, 19
  sensible heat, 17, 19
  specific heat, 14
  temperature, 19
  weight, 19

Stokers, mechanical, 138, 176
  air-supply with, 188
  Coxe, 217
  tests, 178
Straw, 58
Surface-ratio, 149, 191

TABULAR VIEW of results of combustion, 125
Tan, 58
Temperature, boiler in, 151
  coal fire, 151
  effect on action of fan, 257
  ideal, of combustion, 52
  influence on movement of air, 257
  tubes, in, of boiler, 153
Thermal unit, 15
Tube-heating efficiency, 172

ULTIMATE analysis, 69
Unit of evaporation, 94
  of heat, 15
United States Cotton Co. test, 133

VELOCITY-HEAD, 198
  relation to pressure, 195

WATER, 8–18
  bulk, 10
  column due to unbalanced pressure in chimney, 230
  composition, 8, 26
  density, 10
  expansion by heat, 11
  gas, 90
  gauge, 205
  height of column corresponding to pressure, 204
  specific heat, 14
  thermal units in, 16
  weight, 9
Wollaston anemometer, 207
Wood, 58
  composition, 59
  fibre, conversion into coal, 67

ZERO ABSOLUTE, 12

www.ingramcontent.com/pod-product-compliance
Lightning Source LLC
Chambersburg PA
CBHW022028240426
43667CB00042B/1230